IN A DESERT GARDEN

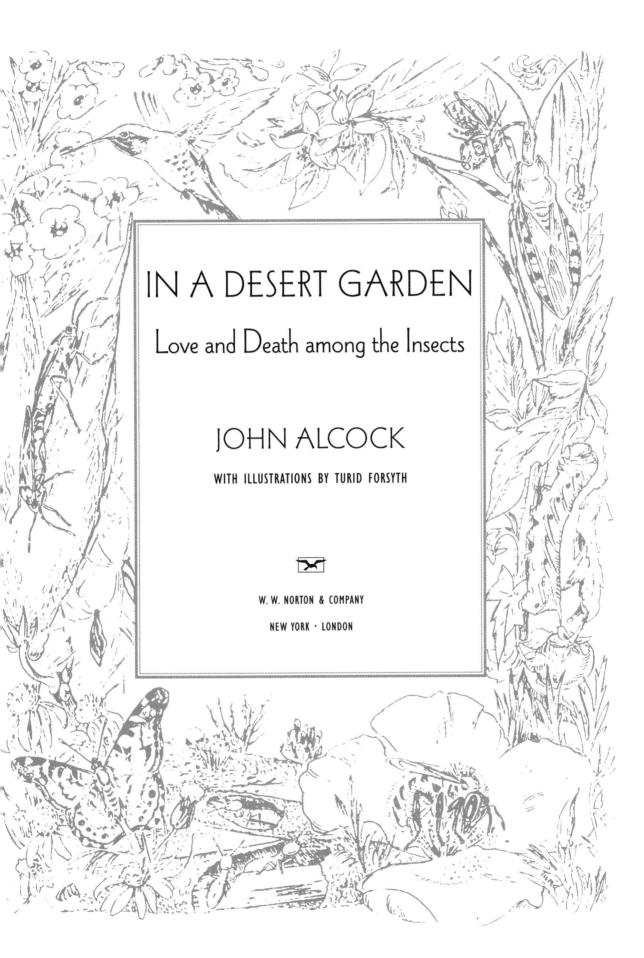

IN A DESERT GARDEN

Love and Death among the Insects

JOHN ALCOCK

WITH ILLUSTRATIONS BY TURID FORSYTH

W. W. NORTON & COMPANY

NEW YORK · LONDON

For information about permission to reproduce selections from this book,
write to Permissions, W. W. Norton & Company, Inc., 500 Fifth Avenue,
New York, NY 10110.

The text of this book is composed in 11/16 Palatino
with the display set in Wade Sans Light
Composition by A. W. Bennett, Inc., Hartland, Vermont
Manufacturing by Quebecor, Kingsport.
Book design by JAM Design

Library of Congress Cataloging-in-Publication Data

Alcock, John, 1942–
In a desert garden : love and death among the insects / by John
Alcock ; with illustrations by Turid Forsyth.
p.   cm.
Includes bibliographical references (p.   ) and index.
ISBN 0-393-04118-2
1. Insects—Arizona—Tempe—Behavior.   2. Desert animals—Arizona—
Tempe—Behavior.   3. Desert gardening—Arizona—Tempe.   4. Desert
ecology—Arizona—Tempe.   5. Garden ecology—Arizona—Tempe.
I. Title.
QL475.A6A44   1997
595.715'09791'73—dc21   97-589
CIP

W. W. Norton & Company, Inc., 500 Fifth Avenue, New York, N.Y. 10110
http://www.wwnorton.com

W. W. Norton & Company Ltd., 10 Coptic Street, London WC1A 1PU

1   2   3   4   5   6   7   8   9   0

CONTENTS

Contents

·

8

PREFACE

Trouble in Paradise, and a Partial Solution

LIVE HAPPILY IN Tempe, Arizona, a suburban appendage of Phoenix, although I take no great pleasure in being part of a megalopolis of more than two million people. No, I grew up in a rural corner of Pennsylvania where the dairy cows greatly outnumbered the people, where you could hear spring peepers trilling in a marsh a mile away on a spring evening, where the dust floated dreamily above the unpaved county roads in summer. The nearest town boasted a few hundred inhabitants, a bridge across a creek, a town hall that burned down when I was a kid, and a tiny general store and radio repair shop. Thus, during my formative years, I became imprinted on solitude, so much so that I experience a vague claustrophobia in places like Phoenix. But my home city has any number of virtues that compensate for its excess of people. Here I speak especially of Arizona State University.

This fine university supplies me with a generous paycheck every two weeks for nine months of the year and gives me the opportunity to regale hundreds of students each semester with monologues on photosynthesis, Mendelian genetics, and Darwin's finches. A tenured college professor, I have more freedom than the average taxpayer would consider good for anyone. To top it off, I belong to a congenial gang of longtime colleagues. We assemble regularly at the Chuck Box over on University Avenue where we compare notes on aging while admiring the beer in our nicely frosted quart-jar mugs.

In addition, because of the quirks of national history, much of Arizona is federally, rather than privately, owned, which means that greater Phoenix floats within a vast ocean of public lands. A forty-minute drive and, presto, I am in the Tonto National Forest where the "forest" is composed of strange and wonderful saguaro cacti and where no one objects when I hop out of the car and set off cross-country on a desert hike of my own invention. Out among the paloverdes and ironwoods, the ratio of cactus wrens to people favors the wrens. The city with its asphalt and whirling traffic seems a million miles away. The country boy in me regains the upper hand for a time.

I love the desert and its restorative powers so much that some years ago I wondered if I might not bring the desert to my home ground instead of always having to make that forty- or fifty-minute drive out to the Usery Mountains or the San Tans when in need of a desert fix. My home ground is small, a standard fifth-of-an-acre lot—9,000 square feet—much of it occupied by the house where I live with my wife, Sue, our two sons having fledged some years ago. The developer or the previous owners ornamented our property with what were the basic landscaping trees of three decades past: an African sumac in one corner of the lot, a line of sour oranges as a hedge along the street that marks the western edge of our property, one lemon tree, and one grapefruit tree. What little that was left of the lot lay beneath a mat of Bermuda grass tufted with unlovable nut grass, forming two beat-up lawns, one in the back of the house, and the other out front.

The more I contemplated my thoroughly ordinary suburban surroundings, the more I longed to trade it in for something that would remind me of what is great about Arizona, its desert habitat. Eventually I did something about it, gradually making over our property by replacing the Bermuda grass with gravel and a diversity of desert plants while installing a vegetable garden out front and compost heap in the back. The virtues of desert reconstruction, front-yard gardening, and backyard composting include the opportunities for weeding, planning, restructuring, harvesting, and compost turning that deflect one's attention from the minor irritants of living in a city. As I transplant baby penstemons from one part of the yard to another, I don't have time to focus on the undercurrent of noise rumbling up from the six lanes of cars rushing or (more often) creeping along the Superstition Freeway a mile south of our home. Turning the compost heap keeps me from reflecting on the fact that I am breathing city air enriched with exhaust from a half million automobiles and trucks. Picking my snow peas, I can always accentuate the positive knowing that I now have the makings of a great stir-fry. A stir-fry will also accommodate some of my Swiss chard, which the garden always produces to excess, or so my boys noted when we lived under the same roof.

Still, it is hard to avoid the feeling of urban encirclement even when one is blessed with a xeriscaped front yard and an abundance of homegrown Swiss chard. The other day I happened to be reading the *Arizona Republic*, a newspaper with an editorial tradition straight out of the Barry Goldwater school of journalism. One of its editorial board writers reflected on the continuing loss of desert habitat as developers rush to meet the needs of the many immigrants to our "paradise." The *Republic*'s editor acknowledged that Phoenix was indeed growing explosively, with no end in sight. Given the attractions of our snow-free environment (incidentally, today it is 109 degrees in the shade), people were going to come in droves. But not to worry; by preserving a desert patch here and there, coupled with intelligent planning of the new communities to be built around these remnants, all would be well twenty, twenty-five, or thirty years down the road. Paradise retained.

Years ago at a political fundraiser for a relatively liberal Republican candidate for the state legislature, I asked whether she would consider actions to reduce the rate of population growth here. I received an answer in the form of a gentle question: How could I, a beneficiary of the joys of desert living, seek to deny to others what I myself had secured by becoming an immigrant to the state? My wife flashed me a nervous look, and I decided to forgo a debate with the candidate, whose views on other matters were considerably more in accord with my own than those of the average Arizonan politician, then and now. But reading the editorial, with its smug assertion that we had urban sprawl and desert degradation under control, rekindled my long-stifled desire to argue with someone about this issue. I raced to my computer and thumped out a letter to the *Arizona Republic*.

I went on for some time about people, freeways, and sewers, and could have continued in this vein, but it was time to get back to the Phoenix Suns–Chicago Bulls ball game on TV. Besides, the *Republic* was right. The good folks from Minnesota and North Dakota are going to keep rolling south to Phoenix. They would keep coming, even if there were dozens, hundreds of us in opposition, each as silver-tongued as the real estate operatives who have sold retirees in Fargo on investing in a townhouse next to a golf course in central Arizona.

Which is why it is an excellent thing to have even a tiny bit of re-created desert in one's front yard as insurance against the day when the real thing will be a two-hour drive down a freeway hemmed in by housing developments of immense scope and numbing uniformity. Moreover, a desert yard and garden provide all those wonderful distractions that help keep one's mind off urban sprawl. I find it both astonishing and gratifying that a fifth of an acre can generate so much time-consuming yard and garden work, as I battle weed invasions, incorporate compost in the garden, and water the vegetables. I don't begrudge my suburban lot any of the time and occasional hard work that it demands if I am to keep it looking vaguely desertified.

Without a doubt, many millions of other Americans share my pleasure in puttering about yards and gardens, but I suspect that most are unaware of a yard and garden entertainment that ranks well above weeding and compost building. I speak of insect watching, an activity that subtropical Phoenix offers in abundance almost year-round. I am one of those moderately unusual (or possibly downright odd) people who observes insects with enthusiasm, to the extent of putting down my rake or fruit-picker to spend some time with a bug. It probably has something to do with my training as a biologist. As a university professor, I teach introductory biology and I also do some research, having for reasons buried deep within my psyche selected insect mating behavior as my special interest.

My enjoyment of insects and enthusiasm for their mating mores stem from an awareness of the delightful strangeness of these beasties and their reproductive tactics. Some twenty million insect species, give or take a million or three, call Mother Earth home. Each of these millions has its own quite wonderful lifestyle. Yet most of us barely acknowledge these creatures, and then only to swat the mosquito on our forehead or to scramble for the spray can to annihilate the paper wasps nesting under the eave by the back door. This book has been written to help readers see another side of insects, namely, their great variety, their nifty ways of avoiding natural predators, their amazing ability to get from there to here, and, especially, their rich and inventive sex lives. I illustrate the delights of insect behavior with examples drawn from the species that live right next to me (and probably you too). The little heroes of this story roam my backyard, and front yard too. Although easily overlooked, once bugs are calmly observed, rather than frantically attacked, they can add a great deal to an otherwise urbanized life. My subjects spice up an environment dominated by concrete; they provide contact with the natural world, much of which we have altered or pushed farther and farther away; they have compelling stories to tell if we are willing to listen to them on their terms.

I begin by describing the conversion of my once insect-unfriendly home ground into a suburban insect oasis, a task accomplished without totally alienating my vertebrate neighbors. The job required changing my standard Bermuda-grass front yard into a patch of reconstructed desert chaparral with a small rectangular vegetable garden set off to the side. This metamorphosis opened the door to a parade of insects who came to stay or

Native bee on desert senna

at least visit for a while, sharing their lives with me while also sharing part of my garden's produce in some cases. After explaining how I transformed a less than satisfactory lawn into an insectarium, I shall in the following chapters wander through the front yard, pointing out the insects on the brittlebush, milkweed, and zucchini while explaining what is intriguing about the brittlebush aphid, the milkweed bug, and the zucchini bee.

How fortunate we are to live on a planet where insects rule supreme. How wonderful to have such a diversity of neighbors, each as fascinating in their own small way as the dinosaur, chimpanzee, porpoise, and other big and obvious animals that provide the raw material for many a television extravaganza. Furthermore, unlike grizzly bears or emperor penguins, our insect neighbors are wonderfully accessible, at times even too accessible. We need not mount an elaborate expedition to deepest anywhere to find an animal whose ways of doing things can astonish and amaze us, fill us with questions, and teach us about the frontiers that exist just outside the front door.

IN A DESERT GARDEN

HOW TO REVOLUTIONIZE
A YARD

Nobody goes there any more; it's too crowded.

—Yogi Berra

HAVE JUST BEEN out to what used to be one of my favorite spots in the desert, a little mountain ridge a half hour from home, more when the Superstition Freeway is not cooperating. The freeway was passable this morning, since I timed my departure to coincide with heavy traffic crawling in as I was heading out. When I arrived at the spot where I usually park the car, a gravel pit used by shooters too cheap to go to one of the regulated shooting ranges around Phoenix, I found the usual eye-boggling carpet of litter covering the entire pit and curling into the nearby desert.

I negotiated my way past the main body of trash and began to ascend the ridge, passing a paloverde that had been cut down by fusillades from shooters using targets on the slope by the tree. A new off-road track violated a nearby side ridge; the last part of the brown gullied track went almost straight up before running out of gas. When I reached the first high point and looked down the other side, I saw that most of the vegetation on the next peak had been burned not too long before. A big black swath of emptiness circled the peak.

I dropped down the ridge slowly to get to the wash below. Fresh tire tracks came right up to the rocks below a little seep. A cupful of water moistened the gravel. Twelve beer bottles, five of them broken, lay at various distances from a fire ring built in the middle of the wash near the seep. By now I was starting to feel just a tad misanthropic.

So I said the hell with it and drove home, jumping out of the car to take a long admiring look at my front-yard oasis. No broken beer bottles there. A good feeling engulfed me, as I congratulated myself for having a place where I can set aside the anxieties of dealing with traffic and overcrowding and litterbugs, a place where I can turn my attention to something I can control, something like yard maintenance.

I used to devote a lot of energy and spare time to mowing, watering, and even edging the bedraggled Bermuda grass that once blanketed our front yard. My lawn, dotted with unwanted weeds and yellowed here and there as a result of erratic watering, was not the envy of the neighborhood. In fact, one neighbor told me gently that only I could top her husband when it came to producing a poster child of a lawn. The front yard did, however, generate enough Bermuda grass to require frequent mowings; I trundled back and forth across the green and yellow sward shoving an aged push-mower in front of me. The mower had come secondhand from a local yard and garden shop, and it tended to lose the right-hand wheel at unexpected moments. No matter how tightly I wrenched the wheel back in place, off it came again in short order.

I cannot remember what source of inspiration freed me from my accursed mower and the lawn it chopped and scalped throughout many a broiling summer. However, one late spring day in 1988 I marched onto the front yard armed with a pump sprayer filled with Roundup at the proper concentration. The active ingredient in Roundup has the vaguely malevolent name *glyphosate.* Suburban users are encouraged not to get the stuff on their skin, in their eyes, much less inhale it. My grass, recently watered, was growing more enthusiastically than usual, the better to absorb the deadly herbicide that I unleashed upon it. After the dousing, almost all of my Bermuda grass turned turtle over the next few days, just as glyphosate's manufacturer had promised. The remaining 10 percent fought a rearguard action requiring another surgical strike some weeks later. I felt not the slightest remorse as I gunned survivors down in their tracks. This was war, war on Bermuda grass and all that it stood for.

Over the rest of that summer I let the chemically scorched yard stand naked under the Arizona sun. No more fertilizer, which had always encouraged the weeds and nut grass more than their Bermudan cousin, whose reaction to my sporadic care and attention had never been as demonstrative as it should have been. No longer would I unkink the hose and spray an insatiably thirsty lawn with thousands of gallons of water siphoned from the Salt River. The river slides out of the White Mountains on the New Mexico border only to die in four man-made reservoirs upstream from Phoenix. From these catchments comes much of the water that keeps our drinking glasses full, our lawns sprinkled, and our cars clean. Now that our lawn remained unsprinkled, our water bills declined. Bits of dead Bermuda grass fragmented and blew out onto sidewalk, driveway, and street. After the end of summer rain, I was thrilled to spot not a single green shoot in the front forty. Neighbors who had long commiserated with me on the tatty nature of my Bermuda grass now looked downright puzzled: what in hell was going on?

I explained that I was about to install desert vegetation in place of the old Bermuda grass, but it would take a while. The installation of cacti and paloverdes was, however, only the last of several time-consuming steps—herbicide application had been only the first. The next stage required rental of a small but jarringly noisy Kubota tractor with a front scoop about three feet wide. Its operation demanded the sort of skills a longtime video player might have acquired. I (a non-videologist) lacked the necessary lever-flipping dexterity, but fortunately the tractor also came equipped with a rear blade that with one shift of a knob fell to the ground with a gratifying clunk. Engaging the clutch, I then persuaded the Kubota to motor ahead. As we inched forward, dirt and dead grass piled up in front of the scraper until such time as I commanded the rear blade to rise, leaving behind the unwanted remnants of a suburban lawn.

Around and around I went, cotton wads stuffed in my ears, scraping and pulling in an effort to slice down through the first six inches or so of "topsoil." Little piles of dirt grew into small mounds. I attempted to lift and move these mounds, using the front scoop with consistent ineptness. Long into the night the struggle went on—man against machine—but by 10 P.M. I had accomplished more or less what I had intended. Something on the order of five tons of dead Bermuda grass and its upper roots and a great deal of clay soil had been scraped up and shifted from one place to another.

In the course of scalping my yard, I had formed a depression that ran diagonally across the lot, a future dry "wash," as dry streambeds are called out West, flanked by ridges of soil on either side. After the tractor rental people retrieved their machine in the morning, I moved into the manual labor mode, using a shovel to refine the crude outlines of the wash and its low banks. This was hard but satisfying work.

Having blocked out the wash, I turned my attention to the western edge of the yard, where I designated an 8-foot × 20-foot rectangle as a future vegetable garden. I covered the rectangle with a 12-inch layer of scraped dirt, first sifting it through hardware cloth in an attempt to remove dead grass roots and lumps of clay. Wheelbarrow load after wheelbarrow load of vegetative miscellany and hard clods made their way to the back alley, to be dumped there unceremoniously for later pickup by the city's hardworking alley-cleaning brigade. The sifted soil remained in the front yard, where it was soaked, dug into the old clay beneath it, and supplemented with huge amounts of commercial soil amendments. The raised beds would, I hoped, provide a congenial home for vegetable plants.

My visions of a vegetable bonanza proved optimistic, to say the least, but with the passage of time, my pocket garden has settled back to the same level as the surrounding yard while providing sufficient Swiss chard to keep me digging and planting year after year as well as offering a vegetative smorgasbord to satisfy any number of insects.

Once my incipient garden stood in place in the fall of 1988, I had plenty of work still ahead on the ex–Bermuda-grass plain before I could boast of having a desert yard. One unfinished step involved placing several tons of fist-sized gray rocks on the floor of the "wash." A local rock and gravel company gladly brought me more than enough of these rounded rocks, which they dumped on the yard. My son Nick and I distributed these materials to their proper positions, a task requiring a stout back and an empty mind. Nick worked on the rockpile with more good humor than the job deserved. Sue had the sense to find something else to do in the confines of the house, but she emerged on occasion to comment on our accomplishments.

Then the rock and gravel company returned to deposit the gravel to go on the yard around the wash and garden. A huge dump truck let fly with many tons of Mission red gravel, quarter inch and finer. A small pyramid soon rested on the concrete slabs of the driveway. Little fracture lines in the concrete skittered out from the gravel mound, testimony to the weight of the load we would have to move.

Nick and I returned to our hard labor, cheerfulness gradually evaporating as the pyramid lost its mass ever so slowly. The wheelbarrow groaned under its burdens. Stout young back and bowed middle-aged back creaked alike. Arms ached. But persistence has its rewards and eventually the driveway regained its usefulness as a place on which to park a car while the yard's gray-brown clay substrate was uniformly concealed under several inches of far more attractive gravel.

As a university professor I have academic friends, who are generous with their advice, dispensed without charge. In discussing the proper method of graveling the yard, some colleagues recommended the interposition of black plastic sheets between clay and gravel. This, however, I did not do, citing the added expense and the aesthetic disaster sure to follow when tattered fragments of black plastic emerged from underneath the ground, like hands from shallow graves, to decay in full view of passersby and so destroy the illusion of desert habitat. The advantage of buried black plastic is, of course, that it thwarts unwanted grasses and weeds from colonizing the gravelly soil. But without these invaders to hunt down and extirpate each spring, what excuse would I have for inspecting each square inch of my creation, gaining and regaining a familiarity with its every feature, sweating the good sweat, smoothing, plucking, pruning, putting my own stamp again and again on my handiwork?

Having skipped the black plastic, I began digging holes about 12 inches wide and 18 inches deep in the buried clay, first scraping to one side the blanket of store-bought gravel. Into these slots I inserted one by one a variety of small desert shrubs and cacti: a couple of brittlebush, three or four penstemons, some fairy dusters, two leafless desert milkweeds, a chuparosa, three globe mallows, several aloes, a desert marigold or two, two creosote bushes, a golden barrel cactus, and two highly branched *Opuntia* cacti, all plants in one-gallon containers purchased at the local Tip Top, or the Desert Botanical Garden's nursery in Phoenix, or the more distant nursery at the Boyce Thompson Arboretum, an hour's drive to the east. Three somewhat larger holes were dug to accommodate two paloverde trees eased out of their five-gallon containers and one fifteen-gallon ironwood tree that I humped over to its designated spot off to one side. Almost all of the plants that I permitted to grace my yard were local Sonoran Desert species. Well, a few (such as the golden barrel cactus) did come from Mexico, and an even fewer, such as the South African aloes and an Australian poverty bush, somehow managed to convince me they belonged despite their un-American provenance.

Although the number of transplants sounds substantial, their planting was actually spread out over many weeks, with additions and replacements right up to the present. Early on, even after several dozen plants had been put in place, the yard still looked empty, quite abandoned, displaying none of the happy clutter of a native patch of Sonoran Desert. Patience would be required.

Some friends and colleagues urged a drip irrigation system for the plants now scattered among the gravel. Once again, frugality won the day. Only plants able to survive without major technological assistance had a place in my front yard. Each of the plants that I ushered into its new hard-packed

Poverty bush
flower

clay home was thoroughly watered on its day of planting and then again at ever greater intervals. Almost all have survived, and most have thrived, confirming the wisdom of my penny-pinching negativity with respect to drip irrigation lines and timers.

As the years have passed, these immature youngsters planted so long ago have grown dramatically, filling in many of the open spaces that appeared nude in the desert yard's babyhood. The yard now has the look of chaparral. To maintain some suggestion of desert, I have on occasion had to dig up and cart away a few original colonists and their offspring. Globe mallow and brittlebush sometimes get carried away with the pleasures of "captivity," taking full advantage of the relative absence of competition from neighboring plants, expanding almost overnight when it rains, so much so that my current policy is never ever to run the hose out to them. Were they to receive extra water, they would balloon completely out of control, sprawling outward, reaching higher and higher, achieving a mass well beyond the restrained dimensions of their true desert cousins.

As my imported plants reached maturity, some proved especially good at producing new seedlings for me to play with. For example, I now know that I can always count on *Penstemon parryi* to generate a small army of baby penstemons each year. This attractive stalked plant with its tubular red flowers yields seeds that clearly vary in what triggers germination. In this they are like the seeds of other penstemons whose germination requirements have been carefully studied. In these other species, some seeds require light to become a seedling, others do not; some need to be chilled before two little leaves pop out of the ground, others do not. As a result, penstemon seedlings appear under a wide range of conditions in both winter and spring. In both seasons, Parry's penstemon do wonderfully well if it rains (or if I sneak some Salt River water to them).

In nature, baby penstemons, brittlebush, and globe mallow all like open, disturbed areas such as the edges of dry desert washes or hillside gullies. This preference for disturbed soil makes them perfect for unnatural yards in a perpetual state of rearrangement. I, for one, relish the opportunities for micromanagement offered by my desert yard. It took me a spring or two but eventually I learned which miniature seedling would grow into a penstemon and which was the product of an unwanted weed, the better to save the former and destroy the latter. I could then transplant baby penstemons

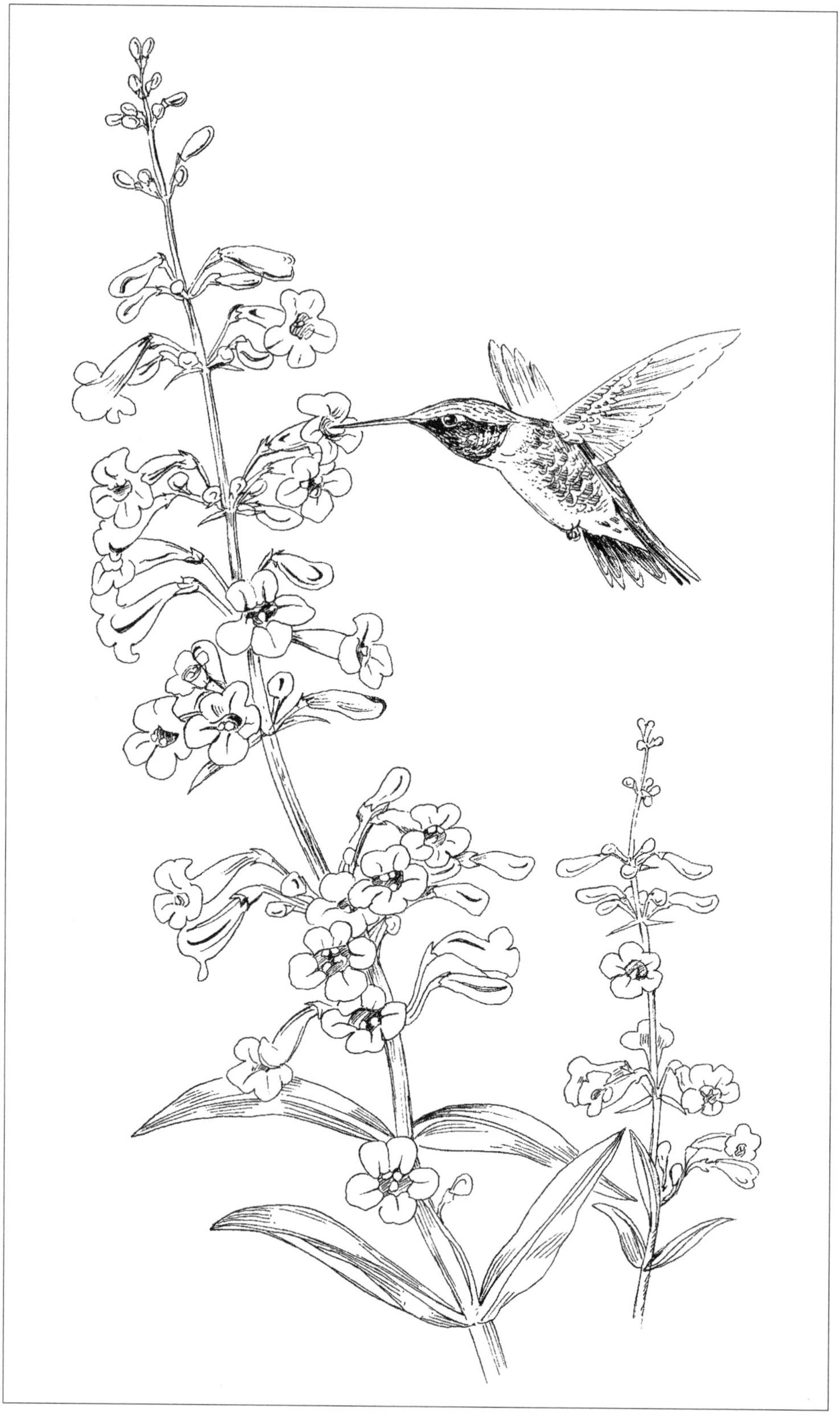

Parry's penstemon with black-chinned hummingbird

in winter, creating small clusters of the plants where I imagined they would flower for maximum effect. In drought years, I have to water the transplants in winter and early spring, but they more than repay my attention with a glorious display of flowers, which attract a bevy of bees to say nothing of Anna's and black-chinned hummingbirds. In March or April, I can count on having a half dozen jet-black carpenter bees and several hummingbirds rocketing from one clump of penstemons to the next in appreciation of my winter manipulations.

The bees and hummingbirds are responsible for pollinating the flowers, so that there will be a new crop of seeds in due course. When researchers have prevented hummingbirds and bees from reaching the flowers of some other penstemons, by enclosing them in various sorts of mesh cages, the plants have not set seed, proof of the importance of these animals to the plants. I do not try to keep pollinators from my plants, and as a result, the penstemons (and other yard plants) invariably produce a bumper crop of seeds—and new generations of seedlings for my experiments in exterior design.

As part of these experiments, I not only foster the growth of favored species, I keep a close rein on those that try to take over too much yard. Since the removal of an overgrown globe mallow merely makes room for one of its youngsters, I need shed few tears for the giant tangle of stems and leaves that I throw on my compost heap, if I am in a composting frame of mind, or cast into the long-suffering back alley, if I am not. The deceased plant will be replaced, sometimes more quickly than I would prefer, by one of its offspring or that of a neighbor.

Therefore, my current yard is always in a state of disequilibrium rather than in the nearly static condition that characterized its Bermuda-grass days. Some shrubs come and go; new individuals arise from seed dropped by adult plants; in the spring, a whole collection of small annual flowering plants, among them desert poppies and Mojave desert bluebells, have their brief moment in the sun before adding seeds to the gravel for storage until the winter rains.

From a botanical perspective, my diversified desert yard is a thousand times more interesting than the grass "desert" it replaced. Moreover, it is an entomological paradise; in sum, the conversion has been richly rewarding. As a result of the much greater variety of plant species in the yard, ranging from trees to desert shrubs and ephemeral annuals to Swiss chard and tomatoes, the specialized food niches available for exploitation by particular insect species have been greatly expanded. Carpenter bees and globe mallow bees, brittlebush aphids and milkweed aphids, these and many other insects can now be accommodated in an enriched environment. I sing the diversity of plants. Down with Bermuda grass! Long live penstemons! Up with insects!

THE GARDENER'S FRIENDS

Am writing an essay on the life-history of insects and have abandoned the idea of writing on "How Cats Spend Their Time."

—W. N. P. BARBELLION

Journal of a Disappointed Man

ARBELLION WAS EXCEPTIONAL in giving insects their due. Most humans care little about bugs, confining their affection to cats and dogs and others of our vertebrate ilk. archy the cockroach once wrote (always in lowercase, even with the assistance of Don Marquis), "The human race little knows all the sadness it causes in the insect world." archy writes truly; the normal run of human being thinks of insects as either insignificant or deserving of annihilation.

True, the balance sheet on insects is complicated. The 100 or so malaria-causing mosquito species are genuine disasters from a human perspective. As many as three million people die of malaria each year; it is sobering to realize that the number of Africans currently infected with the malarial parasite, courtesy of mosquitoes, approximates the population of the United States in 1996. No less disastrous are the crop-consuming locusts and whiteflies and boll weevils and cucumber beetles, which wreak billions of dollars in economic damage and much human hunger in the Third World.

Added to these genuine horrors are the minor ones, including the psychological harm that a suburban gardener must endure when his zucchinis collapse, victims of a lethal squash bug infestation at precisely the moment when his plants were about to produce fruits in a really big way. Likewise when I find a cutworm in *flagrante consumptio* of a once vibrant young vegetable seedling, I feel the heat of vengeance burn within me. Crushing the cutworm between my fingers is the least I can do under the circumstances, and I do it with a clear conscience, indeed with self-righteous enjoyment. The entomologist Howard Evans, usually a great admirer of insects, writes unabashedly of the pleasure to be derived from finding, removing, and stepping firmly upon big tomato hornworms that had been grazing on the family tomato plants before he and his bare feet intervened.

However justified our zeal in destroying malarial mosquitoes, boll weevils, cutworms, and tomato hornworms, most insect biologists would join me in arguing that most insect species deserve neither hatred nor indifference. For example, the great ant biologist E. O. Wilson notes that the total weight of ants worldwide probably equals that of all the human flesh now parading about our planet, and yet all these trillions upon trillions manage

Tomato hornworm

to fit subtly into the world's ecosystems, whereas we make the most dreadful mess of any place unfortunate enough to have our company.

Still, despite the admirable capacity of most insects to coexist with (rather than overwhelm) their environment, despite the wonderful diversity of body shapes and behavioral tricks that insects possess, and despite the awesome complexity and abundance of these little living machines, most people abhor or ignore these marvelous creatures. Even in the field of behavioral biology, my academic specialty, the typical researcher will more quickly embrace a rat or a warthog than watch a fly or a beetle. This pro-mammalian bias deserves a label, perhaps the "Bambi syndrome" or maybe the "baby harp seal factor." Whatever we call it, in fact we usually do find our fellow mammals cuter (even when they are as ugly as a warthog), more understandable, more like us, than the insects that share our earth. The emotionless "face" of a beetle and the inscrutable eyes of a fly leave the average human observer cold, even chilled.

About the only time that people work up even a mild enthusiasm for an insect is when that species happens to labor on our behalf. Consider the honeybee, perhaps the most widely praised of all insects, reluctantly forgiven its painful sting, thanks to the economic and gustatory benefits that it transfers to us via crop pollination and honey. Also included in the select company of "acceptable" insects are the ladybird beetles, lacewings, and praying mantises, on grounds that these predators do good work in agricultural fields and gardens. Under the right conditions, ladybird beetles and their dark speckled larvae will indeed mow their way through one insect pest after another, a fact long recognized by farmers and others. In their book on garden insects, Helen and John Philbrick evaluate the relative merits and demerits of insectkind. They conclude unequivocally that "the Lady Bug larva is a *good* bug." According to some, these insects were once called Our Lady's beetle, on the grounds that they must have been sent by the Virgin Mary to assist gardeners in the war against destructive pests. I wonder who sent us the destructive pests.

It is true that a ladybird beetle, *Rodolia cardinalis*, played heroine in the first modern use of a predatory insect specifically as an alternative to chemical control of an agricultural pest, the cottony-cushion scale. The scale insect had inadvertently been imported from Australia into California, where it found the local citrus orchards a delightful home away from home. Freed from its native predators and parasites, the scale insect took off, and in short order threatened the very existence of California oranges and their growers.

The U.S. Department of Agriculture eventually dispatched an entomologist named Albert Koebele to Australia to search for local consumers of cottony-cushion scale and to bring them back to California. This he did by finding the vedalia beetle, as *R. cardinalis* is known to non–Latin speakers. The beetle does not abound in Australia but Koebele managed to find and ship a total of 524 to California in five installments. When introduced into an infested orchard in Los Angeles in 1889, the scale-munching vedalia beetles found conditions very much to their liking. They proceeded to devour the then abundant prey and multiply with enthusiasm. Growers throughout California learned of this happy event and they visited the trial orchard to collect beetles to take to their own properties. In the space of a year or two, cottony scale was reduced from a major agricultural threat to a minor irritant, thanks to the appetite of an imported ladybird beetle.

The performance of the vedalia beetle so impressed a Florida citrus grower that he hauled some across the country in 1903—even though Florida's orange groves were completely untroubled by cottony-cushion scale. Perhaps this gentleman thought that an ounce of prevention was worth a pound of cure. If so, he was wrong. The vedalia beetle eats almost nothing but scale insects and the imported band of ladybirds therefore traveled with a large quantity of their prey, which kept the beetles alive on

Vedalia beetle eating cottony scale

their cross-continental jaunt. Once in Florida, the beetles were released and some uneaten cottony-cushion scale went right along with them. The beetles promptly died but the scale insects lived on, spawning a new problem that did indeed have to be controlled with another introduction of the beetle, this time under more professional management.

By 1958, the vedalia beetle had become a national asset, exported to fifty-seven countries with cottony-cushion scale problems, and in fifty-five of them the beetle did the trick, thanks to the ease with which it established viable populations when released in a foreign land. For example, from just four individuals grew a mighty population of little predators in Peru. The spectacular success of the vedalia beetle has inspired many other biocontrol programs. Entomologists scoured distant lands for natural enemies of agricultural pests introduced by mischance into places where they were unwelcome. Thus, for example, after a huge scarab beetle, *Oryctes rhinoceros*, made its way to the coconut groves of Fiji and Samoa, the New Zealander Hurbert Simmonds went hunting for a wasp that would parasitize the beetle's larval stage. He hoped that the beetle could be reunited with its consumer, the wasp, which would arrest the destruction of coconut trees by the scarab. Finding a suitable wasp took Simmonds to many an exotic place and afforded him an adventure or two; while rooting around in the dead stump of a palm tree in Malaya, he flushed out a large cobra that "came up between my legs and looked around with expanded hood to see what had disturbed him. The Hindus say that the cobra is a gentleman, and this one after a few moments moved quietly away; so also did I." Simmonds, a man of the stiff-upper-lip school of field work, did not let this potentially traumatic encounter deter him from his task. As a result, he eventually helped establish a scoliid wasp in Fiji that preyed upon the coconut-destroying rhinoceros beetle and reduced the harm the beetle did to coconut commerce.

To be honest, however, this business of introducing a natural predator to control an agricultural pest has produced some spectacular disasters, a prime example being the introduction of the immense marine toad into the canefields of Australia. Someone decided that the marine toad could nicely consume an enemy of sugarcane, another scarab beetle whose larvae develop within the cane stalks much to the detriment of the plant. The marine toad is a large animal with an appetite to match its size; it will eat the scarabs, if they present themselves, but unlike the vedalia beetle and the scoliid parasite of the rhinoceros beetle, both of them specialist predators, cane toads, as they are now called in Australia, do not draw the line at agricultural pests. They will, in fact, eat almost anything that moves. In particular, they have such a perverse liking for other amphibians that a great many of Australia's marvelous native frogs and toads are now endangered.

This downside of biological control, however, rarely shadows the minds of amateur gardeners when they eagerly buy some of our homegrown North American ladybirds, ordering them up by the half-pint from various gardening services. The beetles have been snatched by the bushel from their wintering sites, often located high in the mountains where adult ladybirds gather by the hundreds of thousands before the winter snows come to bury them. When human collectors find these aggregations in the fall, they can sweep up the insects in great quantities for sale to middlemen who eventually ship them to gardeners who think that they will remove unwanted aphids from their home turf. And it is true—many ladybird beetles like nothing better than a steady diet of aphids. However, I suspect that most commercially supplied ladybirds stick around just long enough to regain their bearings after the trials of capture and shipping, then open up their brightly painted wing covers, spread their concealed wings, and fly away in a probably futile search for a more congenial home.

Lacewings are less familiar to most of us than ladybirds, but they too have been pressed into the war against undesirable insects, such as garden aphids and thrips. Despite a delicate appearance and oversized lacy wings, the green lacewing goes after smaller, soft-bodied insects with the resolve of Attila the Hun. I see from my *Smith & Hawken Sourcebook for Gardeners* that I could unleash a horde of lacewings on my garden enemies by purchasing 2,000 lacewing eggs for a mere $19. It strikes me that I would have to have an army of aphids on a very large amount of broccoli and Brussels sprouts to make $19 worth of lacewings a wise investment. Moreover, Smith & Hawken do not guarantee that all 2,000 hatchlings will reach adulthood, nor do they speculate on how many of those that do become full-sized lacewings will do so in the garden of the paying customer, rather than moving on to assist a nonpaying neighbor.

Still, hope springs eternal and the idea of letting nature control nature, and thereby avoiding the use of pesticides on one's vegetables, is a laudable one. If commercially supplied ladybird beetles and lacewings won't do the trick, what about praying mantises? These large insects are also notable killers in their own right, and they too have been chosen as potential friends of farmers. In fact, an attempt at the biological control of insects resulted in the importation of Chinese praying mantises in 1896 for general release in the United States. Even today, Smith & Hawken, insect purveyors, will ship me or you praying mantis egg cases (some almost certainly produced by descendants of the imported Chinese variety). The cost for three cases is $9, making the mantid deal substantially cheaper than the lacewing option. On the other hand, the three egg cases will yield only 600 baby mantises, according to Smith & Hawken, whereas I could secure the services of 2,000 lacewings for $19. It's a tough call. I am assured by Smith & Hawken that when the mob of nymphal mantids emerge from their egg case, they will "devour almost any insect in their path."

This bit of hyperbole collides head-on with a report in the *Journal of Insect Physiology* on the development of prey catching in the Chinese praying mantis. The authors state that day-old mantises are quite unable to catch prey, as their binocular vision is still coming on line at this time. On the second day of life, baby mantises of this species manage only "timid, unaimed strikes," and it is not until the third day that they are beginning to turn into predatory terrors of the sort to warm the heart of a Smith & Hawken customer.

It is true that a three-day-old mantis has many days of insect catching ahead of it, if all goes well. Members of the introduced Chinese mantid, *Tenodera sinensis*, hatch out in April in the eastern United States, reach

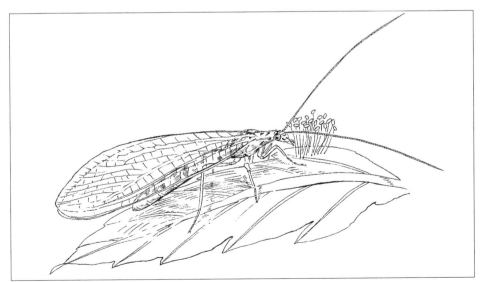

Lacewing, a delicate predator

adulthood in August or September, and live until the first killing frost in October, by which time they will have sent many fellow insects to their ultimate reward. Females would be the better investment for the garden, since, in some species at least, they live longer as adults than males, around ninety days instead of forty in one well-studied African mantis.

Praying mantises of both sexes come equipped with spiny nutcracker-like forelegs, great for grasping flies and the like. Once captured, prey are pulled back to within comfortable grazing distance. The time required for a strike and capture has been recorded on high-speed film at 50 to 70 milliseconds. That's fast, and it has to be because a blowfly can take off within 45 to 65 milliseconds after being disturbed by a moving object. Those of us who have tried to snatch a perched fly from our school desks or a pant leg know that a reaction time of 45 to 65 milliseconds is usually good enough to leave us empty-fisted and mildly disappointed. Luckily for us, our survival does not depend on catching flies with our hands. If we were up to praying mantis standards, however, we could nail flies eight or nine times out of ten tries, a really impressive achievement by schoolboy standards.

But despite the blinding speed of their attacks, and their large size and healthy appetites, I know of no research showing that mantises remove more than a tiny fraction of the pests that descend upon backyard gardens and sweeping fields of grain. The few cases of successful biological control resemble the vedalia beetle story, in that they generally involve highly selective insect predators or tiny parasites. If the preferred victim or host is a pest species, and if the predators or parasites can become numerous enough, then a fairly high proportion of the unwanted pests wind up in the guts of their specialist enemy. Unhappily, success stories in this field are rare.

So if you have around your garden a couple of three-inch-long Chinese mantids or several of the smaller mantids native to North America, your insect troubles probably are not over. Still, you can take satisfaction in their presence for any of a number of reasons, aesthetic or intellectual. Each time I see a mantis in my parents' Virginia garden (unhappily, I do not have them on my home ground), I am reminded of the vigorous and continuing debate that has raged over the nature of mantid sexual behavior. Academicians enjoy arguments, debates, and disputations to an extent that would cause concern were these arguments about matters of import. Fortunately, most biological debates revolve around less than earth-shaking matters, as the case of the "mantis wars" will demonstrate.

This issue so stirring to the passions of my colleagues has to do with "sexual cannibalism" in the mantids. The story that male mantids are usually eaten by their sexual partners has achieved widespread distribution. Thus, Helen and John Philbrick report discreetly that the mantis "is so spectacular that he, or rather she, (because the female devours the male at a certain stage of their life cycle) has always received a good deal of enviable publicity."

Whether enviable or not, the publicity was there all right, probably arising in part from research on mantis sex in the 1930s by Kenneth Roeder of Tufts University. Roeder was a wonderfully creative neurobiologist interested in how the nervous system of insects regulated their behavior. In the course of his work on mantis mating, Roeder found that he was more likely to get a captive male to copulate if he first removed the male's head before putting him next to a receptive female. The headless male was perfectly capable of moving about in a circle until he touched the female, at which point he would mount her and copulate competently, with satisfactory sperm transfer.

From this work, Roeder concluded that the copulatory behavior of the mantis was normally inhibited by signals coming from nerve cells in the male's head. Removal of the head eliminated neural inhibition and so facilitated a response that was controlled by other components of the nervous system, which are located in the male's thorax and abdomen. Probably as a result of these experiments, the tale grew that males *had* to have their heads removed if they were to copulate, and that they therefore cooperated in their decapitation by lusty females.

If it were true that male mantids really must die in order to reproduce, it would definitely be a surprising and counterintuitive phenomenon worth getting worked up about. One would want to know why male mantids had not evolved a different kind of neuronal system, wiring that would not order them to commit sexual suicide when mating with and inseminating their partners. If a nonsuicidal male appeared in a given species of mantid, would not this individual experience much greater reproductive success than his suicidal fellows? He would have a chance to mate with many females, rather than expiring in the jaws of partner number one. He ought

*Male praying
mantis copulating
with and without
his head*

to leave a lot of descendants, setting in motion a chain of events that would over time lead to the elimination of sexual suicide and the neuronal mechanism that once caused it to occur.

But wait a minute (said a few evolutionary biologists). What if the male's sacrifice of himself provided his partner with a banquet of protein, calories, and nutrients that enabled her to go on an offspring-producing binge? If these extra mantis babies carried the genes of their cannibalized father, he would gain descendants by virtue of his readiness to provide a protein-rich meal to his partner. These added offspring might compensate their father for the loss of additional opportunities to have young by mating with other females, especially if these other opportunities were few and far between.

The idea sounded plausible to some, but less plausible to others, including W. Jack Davis and Eckhardt Liske, who wondered whether sexual cannibalism in mantises might not be a total nonphenomenon, a pure artifact of laboratory conditions. In their own laboratory studies of the Chinese praying mantis, Davis and Liske found that males almost never died from attacks by females, partly because they approached potential mates extremely cautiously. If detected, mantis males sometimes used a variety of courtship displays at a distance to avoid getting too close too quickly. If undetected, however, some males succeeded in creeping up to apparently unsuspecting females, leaping on their backs, and copulating without giving their partners a chance to dine upon their heads first.

Davis and Liske concluded that the extent of female consumption of males during copulation had been greatly exaggerated. Males gave every appearance of trying to avoid sexual suicide, rather than throwing themselves into the jaws of a mate. Moreover, males were entirely capable of mating with brain intact, as shown by Davis and Liske's many observations of mating pairs in which the male retained his head from beginning to end of the nuptial exercise.

Word of Davis and Liske's research somehow filtered out to the nonacademic world, including the schedulers at National Public Radio. An NPR interviewer contacted Professor Davis, who told his listening audience what he and his colleague had discovered. He concluded that many of his fellow academics had proven remarkably gullible in buying the accounts of headless copulators and sexual suicide in the mantis.

However, some copulating females of some mantid species (there are on the order of 2,000 described species) clearly practice cannibalism under natural conditions, with the female proceeding to consume her mate slowly, head first, thorax next, finishing up with dessert, the plump abdomen. The entomologist E. S. Ross possesses a fine photograph taken in the field of a female mantid in the act of devouring her copulatory partner. Even Davis and Liske admitted that they had seen *hungry* females in their laboratory trying to kill approaching males, and sometimes succeeding.

Several careful studies of mantises living in captivity or under natural conditions have appeared since Davis and Liske's work. An interesting report comes from S. E. Lawrence on a free-living British population of *Mantis religiosa*. Lawrence discovered that, even outside the confines of a laboratory, these mantises have a predilection for cannibalism while mating, which occurred in about a third of the pairs she observed. In addition, Larry Hurd and his crew of Maryland mantid watchers calculated that males in one natural population of the Chinese mantid ran a 17 percent chance of being killed and eaten by a female on any given encounter. These findings put to rest the possibility that cannibalism by female mantises is *purely* a laboratory-induced artifact.

Without doubt some male mantises are not eager to die in the arms of a possible mate, nor do they have to have their head removed in order to overcome certain sexual inhibitions. However, the combination of the male's ability to reproduce while headless and the female mantid's potential to convert a male into supper leaves enough of the mantis "myth" intact to provide a phenomenon worth arguing about: namely, does the male mantis, as well as the female, benefit from copulatory cannibalism, *when it occurs*, with the male leaving more descendants by virtue of raising his partner's fecundity on those (rare?) occasions when he becomes a nuptial meal?

The fact that male mantises in the field or laboratory generally approach females with extreme caution, freezing whenever their would-be partner looks their way, suggests that females are dangerous, and that males are "trying" to avoid being eaten. These findings jibe with the argument that cannibalism is adaptive for females but not for males. However, Dr. Lawrence concludes that it is still possible that males of "her" species could gain by being eaten in the latter phases of the reproductive season. As the summer progresses, a higher and higher proportion of the female population has mated, judging from the presence of sperm in their sperm storage organ. Although some females evidently mate more than once, it is probable (but not certain) that females that have already acquired sperm are more reluctant to mate than virgins. If so, a male's chances of mating decline over the course of the season. There could come a time when his chances of finding an additional receptive female are so small that he can gain more by permitting a female cannibal to end his life rather than living to search fruitlessly for mates another day.

This argument may not be a wild-eyed speculation. Consider the Australian redback spider. The redback is closely related to our black widow spider, a native of Arizona and once a regular part of our lot's fauna, although in recent years it has become extinct around the house for unknown reasons. Females of both spiders are notorious for their ability to inflict venomous bites on unlucky humans who contact them. Males of both species are tiny, weighing a mere 1 to 2 percent of the average female, and they pose no risk to human health and welfare. But they have their own small claim to fame—like certain male mantids, males of both spiders sometimes make copulation the last thing they do in life.

In order to copulate, male redback spiders first take up residence in a female's web, there to wait for a propitious moment in which to approach their would-be mate. The approach is made cautiously while the male determines the sexual receptivity of his giant web partner. But if she is in a good mood, he will be able to position himself in such a way as to insert

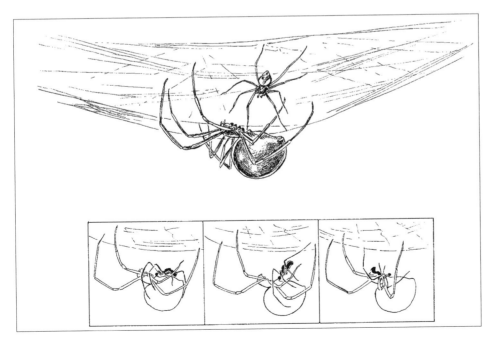

Redback spiders mating

sperm into her genital opening with special leg-like structures called pedi-palps. The male first deposits sperm into a pedipalp and then reaches over to put the sperm into the appropriate part of the female, repeating the process to make a series of such transfers. After some time spent moving sperm about, the redback male may perform a special somersault that puts his small body directly in line with the female's large jaws. She may respond by using those jaws to kill and consume him. His small size means that his nutrient contribution to his mate is negligible but it still takes several minutes to polish him off, and these moments, although hard on the male, are extremely important to his paternal legacy. Males that are eaten fertilize far more of their partner's eggs than males that forgo sexual suicide. When females don't eat their partners, they are likely to mate again with another male, reducing the proportion of eggs that their first mate gets to fertilize.

The real puzzle here is why all males don't throw themselves into the jaws of a mate, since male redbacks live only a few weeks as adults in any event and those that try to make it from one web to another usually die en route. Under these circumstances, there is little to be gained from staying alive to try for another mating with another female, and much to be derived from occupying a mate in terms of fertilizing more of her clutch. The study of redbacks continues, but already we know that researchers should consider whether eaten mantis males give their sperm an advantage in fertilizing their cannibal partner's eggs. If so, maybe some males under some circumstances have more to gain by facilitating their demise than by trying to avoid it.

In fairness, the evidence from mantis studies to date points strongly toward the hypothesis that copulatory cannibalism benefits only females. The data do not favor the alternative view that males sometimes adaptively commit sexual suicide in order to feed their mates or to give their sperm an edge in the race to fertilize a female's eggs. Still, I for one am prepared to wait for more information. Why rush to judgment here? Mantises have been around for millions of years and it may take us a few decades more before we can fully resolve the challenge posed by the occasional occurrence of cannibalism by mating females. While we work on the mystery, may the survivors do what they can on our behalf in the garden, waiting patiently in a prayerful mode for a pest to come their way.

COMPOST LOVERS

To what purpose is this waste?

—MATTHEW 26:8

O H, THE SMELL of compost. What a wonderful primeval odor, especially when combined with the heat of the heap, which warms the gardener's soul; trash transforming itself into the essence of soil. Best of all, it's free—if you can convince yourself that the hours of labor involved in compost building and turning have no cost. You may also have to overlook the admittedly modest expense of a sack or two of commercial compost, since compost piles respond nicely to microbial inoculations.

My recipe for composting in the desert involves more than the infusion of the occasional two cubic feet of store-bought composted sawdust into a mound of mulberry leaves, citrus clippings, grapefruit and orange rinds, discarded brittlebush and globe mallow, snow pea vines after the harvest is over, grasses pulled from the desert yard where they are not welcome, and a wide variety of miscellany and etcetera. My parents' Virginia garden boasts a splendid mound of compost, piled up over an area 8 feet square, down only about a foot or so into the ground. In contrast, my desert heap, less grand by far, is installed in a shallow grave somewhat more than 2 feet deep, and only about 3 feet wide but 10 feet long. In Virginia, frequent rains keep the compost wet, sometimes too wet, and so a deep, moisture-retaining pit is not needed there. In the desert, however, a deeper trench helps keep the composted material from drying out, a goal that I also advance by covering the mound with a rectangle of old rug or an abandoned window shade.

When I began my career as an Arizonan composter, I did not dig a pit but instead piled the materials onto an open patch of ground. Then, in order to keep the moisture level up, I hosed the heap down regularly with water from an outside tap. This procedure yielded mature compost in due course, but when I mulched my vegetables with the stuff, I noticed a white salty rime on the material as it dried out. Eventually I deduced that by using Salt River municipal water to keep the compost heap cooking, I was in effect salting the heap, negating its value as a soil amendment.

My new and improved system of composting focuses on retaining the moisture contained within the plant material tossed into the trench. However, I do also collect rainwater as it sluices off the roof of our house, on those rare occasions when it actually rains in Tempe. Afterward, I stagger around the house carrying some of this presumably salt-free water to the compost heap, if it seems to be drying out. The results of this method have been gratifying. At least I am no longer poisoning my vegetables with salt when I dump a wheelbarrow load of dark brown, friable compost on the garden.

I have learned other things on the road to becoming the compleat compostmeister. Early in my gardening career I wondered what I could do to add to the rather modest supply of vegetable waste generated by our household and surrounding yard. For much of the year, my compost heap seemed underfed. In a moment of inspiration, however, I realized that I had access to a nearly inexhaustible supply of the raw materials for compost right in the Zoology Department. At the time, our Life Sciences Building contained an animal holding facility where rats and mice frolicked. Rats and mice generate great quantities of urine-soaked wood chips sprinkled liberally with fecal pellets. Wonderful, I thought. Cellulose plus nitrogenous wastes. The perfect combination to get the compost really cooking.

So it was that I stuffed three huge plastic sacks of used rat and mouse bedding into my Volkswagen Beetle on the afternoon that Larry, the technician in charge of the facility, designated as pickup day. Returning home, full of hope and enthusiasm, I dragged the bags of urine, feces, and wood shavings around to the backyard compost processing center. As I dumped the future compost onto the pile, it did strike me that the three monster bags may have been too much of a good thing. The resulting pyramid of would-be compost also impressed my wife, who took the time to relay some of her misgivings to me.

But since I wasn't about to put the mouse droppings back in the plastic bags, I decided to persevere, confident that nature would work miracles in due course. To get things going, I hosed down the pile and sat back to wait. Prior to even one small miracle, however, something else occurred. The mound heated and stewed away, producing a ghastly organic stench. The smell of fermenting mouse dung and urine filled the backyard, overflowed the fence, and swept down the dirt alley like a wave of ammonia from a wrecked freight train. I have no doubt that you will believe me when I say it was a most unpleasant experience for me, my loved ones, and my neighbors.

My loved ones knew the culprit's identity. My neighbors did not. I overheard two of them talking, wondering why on earth the Dumpster in the back alley had become so horribly smelly of late. I left them in ignorance.

After several weeks my backyard experiment stopped stinking. It was months, however, before I had usable compost, but happily long before then the air had returned more or less to normal in our immediate neighborhood. I told Sue that one such experiment was enough. She congratulated me on my ability to learn from experience.

In more recent years, my revised method of composting has produced, happily, a steady flow of pleasantly aromatic material. I screen the maturing compost, collecting what passes through the quarter-inch mesh of a large square of hardware cloth, returning the undigested larger pieces back onto the pile. As I shovel, shake, and bag the earthy humus, I have many opportunities to encounter invertebrates that are as enthusiastic about compost heaps as I am, if not more so. Small yellow ants race about on the screened brown-black product, their societies totally disrupted by the tossing and turning of their home. Pillbugs wander over the processed compost, unfailingly befuddled, while the sleek enameled earwigs glide over the surface like miniature snakes as they search for escape routes out of the wheelbarrow that holds the filtered compost. Those that climb up and over the lip of the wheelbarrow waste no time returning to dark tunnels in the half-decayed plant matter piling up in the compost grave.

Earwigs are low-slung creatures with an aversion to light. As a result, they are almost never found out in the open during daytime, but instead stay underground, slipping into narrow crevices, openings under rocks, and the interstices of compost heaps, when available. Their fondness for dark retreats may account for their common name, which is derived from the old wives' tale that they like to crawl into a sleeping person's ear, a story that entomology textbooks nervously debunk.

But earwigs are so strange-looking that they ought to have some unspeakable practices, even if they do not insert themselves into human ears. Long and thin with wing covers that only come partway down their back, they are endowed with two large curved pincers that project out far behind the animal's naked abdomen. The pincers make earwigs look like some sort of arcane tool in miniature, perhaps a novel device for opening tiny vials.

The fact that earwigs are out of sight during the day means that the relatively few people who wish to observe them have to become nocturnal entomologists. My nighttime encounters with earwigs have occurred when I have gone on the prowl after dark in the garden. I do not often garden at night, but there are times when it is essential, particularly when the cricket population has burgeoned in late September or early October.

September is the cruelest month in central Arizona, the fifth month in a row of blindingly hot days and insufferably hot nights. October, on the other hand, has some small redeeming virtues. It is the month when a modest amount of sanity reasserts itself, with daytime highs falling below 100°F, praise the weatherman. After the preceding months of hot weather torture, something in the 90° range stimulates deep relief and profound gratitude in all but the most heat-addled desert rat. Because October offers an Arizonan version of temperate daytime temperatures, it is a time for planting the winter garden, with an emphasis on snow peas. Even in the small area available to me I try to squeeze in some chunky sugar snaps, some mammoth melting snow peas with their huge pods, as well as some reliable dwarf gray sugars. If I can get the peas well established by late November, they will endure the coldest part of the year quietly, ready to take off when spring comes with longer days and warmer temperatures, at which time a bonanza of highly edible pods awaits the well-organized gardener.

I may dream of mature snow peas; crickets, however, find seedling peas poking tentatively through the warm October soil an irresistibly succulent temptation. After an epicurean cricket has visited a row of baby snow peas, the wilted leaves of fallen seedlings, once pregnant with promise, lie in a line, like West Point cadets with heat stroke.

The next evening I make the rounds, armed with a flyswatter and firm resolve to practice insect control the organic gardener's way. I scan the garden for my orthopteran enemies, guided by a flashlight with some assistance from a nearby streetlamp. The slowly moving flashlight beam transfixes a cricket on a pea-killing mission; I raise the flyswatter, inch forward, and—strike! With luck, I will send the herbivorous insect to its reward while missing the plant that I wish to rescue.

This labor-intensive, chemical-free solution to cricket depredation has its limitations—sometimes the cricket sidesteps the descending swatter, leading me to crush a pea seedling instead. Still the technique generally produces a gratifying result, and there are bonuses too, when, for example, I find a juicy cutworm perched out in the open on the edge of a Swiss chard leaf. During the day, cutworm moth larvae retreat either into the heart of a well-established clump of chard or into soil litter some distance from their food source, making them difficult to find. At night, when they come out to join the crickets, taking food out of our mouths, they can be detected much more easily and firmly dispatched.

Sometimes during these utilitarian search-and-destroy missions, I have stumbled on an earwig or two, which scuttled off when exposed by the flashlight. But once an earwig so spotlighted remained where it was, permitting me to see in its pincers a chunky fly, perhaps a dead one, which it had scavenged from somewhere in the garden. Or did it find and kill a sleeping fly? Earwigs are said to be omnivores; some species are known to feed on a wide variety of animal and vegetable matter (but not on earwax, thank goodness).

Few published reports tell of earwigs using their pincers to capture or transport prey, but many more describe the pincers' role in combating predators. Earwigs can twist their flexible abdomen about with the greatest of ease, pinching the enemy with these sturdy weapons. Thomas Eisner, a world expert on insect defenses, drew on personal experience when he reported that some earwigs can draw blood with these devices.

The European earwig, *Forficula auricularia*, supplements its mechanical defenses with the forceful secretion of a liquid containing quinones, a chemical that ants, for example, find highly repellent. Two pairs of glands containing the quinones are located on the anterior portion of the abdomen. By twisting its maneuverable abdomen about, an earwig pinched on the right midleg or the left hindleg by a biting ant can fire droplets of quinones at the right midleg or left hindleg, as needed. Ant muggers doused with this earwiggian version of Mace retreat the worse for wear. Persons triggering the chemical response, as when attempting to remove or kill an unwelcome earwig in their house, will also find it disgustingly odoriferous.

Returning to the subject of pincers, it is more than mildly intriguing that males of the European earwig and some other species have larger and somewhat more powerfully built pincers than the corresponding structures on females. This is true even though female earwigs are generally the larger of the two sexes. There is a class of biologist, to which I belong, whose eyes light up when they hear that the males of a species are endowed with special devices that might be weapons and that either are not possessed by females or at least are reduced in size in the gentler sex. This information signals a species whose males probably spend a good deal of time employing these weapons on one another.

A wonderful menagerie of insects exists whose males have horns, or antlers, or immense jaws, or enlarged and spiny hindlegs. In almost every case, observation of males socializing with other males reveals that they use these structures when fighting each other. So, for example, the elaborate rhinoceros horns of certain scarab beetles, which Darwin guessed might serve to attract mates, actually turn out to be used in jousts between competitors, with males of some species using their horns in conjunction with a head toss to flip opponents upside down or off perches. Or take the truly bizarre antlered flies of New Guinea and northern Australia. The males of one species have protuberances growing out of the sides of their heads, creating structures that look for all the world like a fly's version of moose antlers. When two rivals meet, they go at each other as if they were indeed miniature moose, facing off head-to-head, clashing antlers, pushing and shoving until one male turns abdomen and runs.

And what are the rhinoceros beetles and antlered flies and kicking plant bugs worked up about? It's not subtle. The big boys are fighting over females and the chance to mate with them. The struggles between males typically take place in the company of females with the winner left alone to mate with them, all of them, if he can stay in charge long enough.

So the fact that males of some earwigs have bigger cerci or forceps, as the pincers are variously called, might be the evolutionary result of fights over females between males in the past that were generally won by individuals that happened to have the more powerful and painfully effective pincers. If the big-pincered males mated with more females and had more descendants that carried their genetic information, then today's male earwigs would develop the attributes of past winners—namely, the big pincers of their combative ancestors.

This possibility is plausible, but it would be good to check the hypothe-

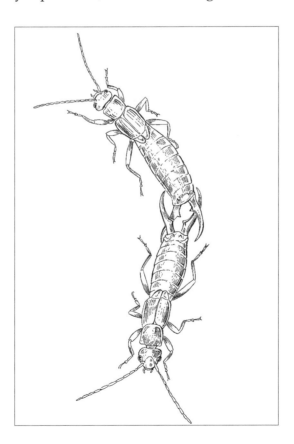

Fighting earwig
males

sis against reality, which has been done, albeit only for a handful of earwigs (of which there are more than 1,700 described species). The European earwig is well studied, thanks primarily to its worldwide abundance, courtesy of cargo shippers who have accidentally distributed this stowaway species to almost all the continents of the world. The first European earwig to set its six feet on the soil of the United States had done so by 1907 when one was detected in Seattle. The earwigs in my garden may well be this species, which was found in Arizona several decades ago.

In any event, males of the European earwig seek out sexually receptive females in their nest burrows, which are often placed under rocks. The nests are built to accommodate not just the mother, but also her young when they hatch from eggs that the mother lays. After mating and laying her clutch of eggs, the female seals herself inside the chamber where she remains without food until the young have hatched. She then opens the burrow and will forage at night by herself, collecting food that she may bring back to the nest to share with her brood.

After a number of days, her youngsters join her on these nocturnal rambles, returning together to the safety of the nest after a bout of foraging. To facilitate their reunion at the burrow, earwigs have scent glands on their "feet," which leave an "odor trail" that family members apparently can recognize and use to relocate their burrow and identify one another. This ability is particularly important to the youngsters, since they are likely to be eaten by adult strangers, a severe lesson in family values.

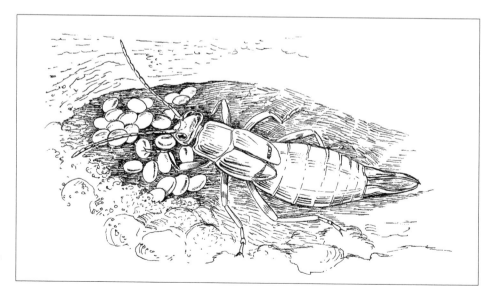

*Earwig mother
with young in nest*

Males take no part in parental care. Their primary mission in life is to encounter willing females that have not yet laid their eggs. When a male finds an unattended female, he remains with her, copulating at intervals, the better to fertilize all forty to sixty eggs that his partner will eventually lay. After the clutch is complete, females drive their male friends away, sealing the nest burrow for the "incubation" phase.

During the period when males occupy burrows conjointly with their mates, they sometimes have to contend with wandering males, which may try to insinuate themselves into the nests. If a wandering male can evict his rival, he will then mate with the female and fertilize some or all of her eggs.

In the laboratory, if two male earwigs occupy the same container with a female, fights usually break out between the males. The ease with which male aggression can be elicited has enabled researchers to explore what it takes to be a winner in the European earwig. These entomologists have exploited the fact that males of this species come in two categories with respect to forceps size, large and small. Two males can have the same body mass and yet differ greatly in the size of their abdominal pincers. When two males of the same body size but unequal forceps are paired off, males with the larger weapons usually force opponents with smaller ones to retreat quickly, even if it means relinquishing a mate to their better-endowed rival.

It looks like big forceps help males dominate rivals and gain mates. If so, why do males with relatively small forceps persist in the European earwig and some other species as well? The answer to this question remains elusive, but the solution to the puzzle may lie within the recesses of a compost heap somewhere. It could be that there are many earwig populations in which the density of rivals is low enough that males with small pincers are rarely challenged by opponents with larger pincers. Under these conditions, male earwigs that took the time and spent the energy needed to develop a Schwarzenegger physique would gain little reproductive return from their investment. As a result, males with the hereditary disposition for smaller, cheaper-to-produce cerci will have opportunities to sow their wild oats and their special genes at the same time.

Or it could be that males with large and small forceps have the same genetic information for forceps development, but some individuals get lucky, have a lot of food as they are maturing, and can afford to produce an impressive set of weapons, whereas other males, through the luck of the draw, grow up in an environment with less food and therefore have to settle for minor cerci. If brother earwigs were separated in their early stages and fed different amounts of food by a careful experimenter, the appearance of adults with major forceps only in the well-fed category would be powerful evidence in support of this hypothesis. My compost could supply the mother earwigs required for such an experiment. You need only ask.

IN PURSUIT OF earwigs, we have wandered far from the other insects that call compost "home." Needless to say, a precise list of heap inhabitants varies from season to season and locale to locale, a point clear to me when I visit my parents' home in northern Virginia. With their many acres, their cowpie-splattered fields, and large, weed-generating gardens, my parents have the room and wherewithal for a truly impressive compost heap. Their eagerness, however, to build and turn the mound is modest, and so I generally assume the duty of compost engineer on my visits. To this end, I first dig out the old compost, working my way down to the harsh red clay underneath the black and brown organic residue of quadrillions of microbes that have done their magic on what were once corn stalks, chickweed, pigweed, cantaloupe rinds, and steer pats shoveled up one by one from the pastures and transported to the pile, where they intermingled with coffee grounds and old hay. Sweat pours down my face, trickling from forehead to the end of my nose, dripping from eyebrows to eyeglasses, drenching a mud-stained T-shirt. I fill the wheelbarrow with the first load of mature compost, and trundle off to offer the stuff to my parents. They are kind enough to admire my present, before permitting me to spread it on the flower borders or to add it to the mini raised beds where they grow lettuce and basil.

The satisfaction of this activity fresh in mind, I return to the now empty shallow pit. I begin to fill the depression with a considerable quantity of garden weeds and tall half-brown grasses provided by my father when he mowed the adjacent slope with a tractor-pulled weed hog. The ancient John Deere tractor spouted fumes and smoke, which floated back in the driver's face, but it got the job done, enabling me to rake up a substantial quantity of raw materials for my operation.

Once I have a foundation of fresh (that is, uncomposted) vegetation filling the pit and rising upward two feet or so, I toss on several shovelfuls of unreconstructed clay and incompletely digested compost set aside when the old mound was dug out. Later today, tomorrow, or the next day I will add many more weeds and hay as well as some cowpats, ideally moderately moist ones, to supply nitrogen for the microbes who would like to multiply in the environment I am creating. For the moment, however, I plan to content myself by adding the kitchen scrap heap to the incipient compost pile.

My parents discard melon rinds, coffee grounds, scraps of tomatoes, old green beans, oversized cucumbers, and all other vegetarian castoffs from the kitchen at a place on the edge of the garden, closer to the kitchen than the site where I hope to construct my ceremonial pyramid of refuse. I shall move these generally well-rotted materials from their initial resting place to my growing compost heap.

But before I do, I admire the small army and air force of insects assembled on or above the effluvium of the kitchen. Bald-faced hornets buzz in to snip and sip at a rotten peach. In this activity they are joined by an even more intimidating wasp, a European import that has in recent years spread into the countryside, a gargantuan monster that looks like a yellow jacket bloated fivefold in a mad scientist's laboratory. I keep one eye on the European wasp; no sense irritating it for the thing may not recognize or tolerate the entomologist in me.

I place my other eye on the layered strata of vegetable decay. Several species of small and delicate flies perch on or hover above the residue. I know little or nothing about them; I would be surprised to learn that anyone knew a great deal about the natural history of even one of these insects, with the exception of fruit flies belonging to the species *Drosophila melanogaster*. This fruit fly is a cosmopolitan compost insect and one of the most intensively studied animals on earth, owing to its usefulness in the study of genetics.

A fruit fly walks on the surface of the mound. The flylet pauses to lower its proboscis onto moist exudate of some sort oozing from a decaying green bean. While feeding, it fails to notice a much larger, dark brown beetle creeping stealthily toward it. Slipping among the jumble of discarded beans, the beetle never loses visual contact with the fly, which shifts position slightly but then continues to suck up fluids. The creeping beetle looks for all the world like a domestic cat stalking an English sparrow distracted by food on the ground.

The beetle inches forward, then pauses, and suddenly bursts ahead directly at the fly, which fails to react in time. In a fraction of a second, the

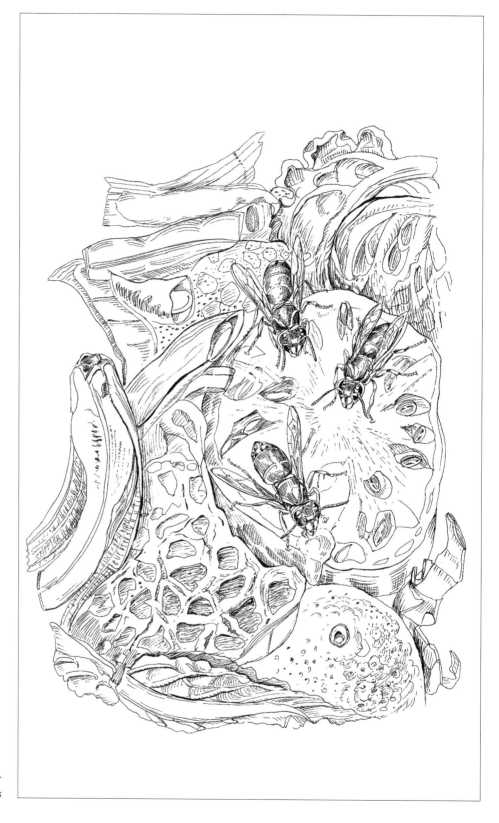

*Compost with bald-
faced hornets*

fly is grasped by the formidable curved jaws of the beetle. The killer impassively crushes and crumbles its victim into bits, consuming the soft body parts before discarding a bolus of inedible cuticle, which falls onto the compost mound where it will be broken down eventually by bacteria and fungi.

The fly-eating beetle belongs to the family Staphylinidae, a fact that it announces to the world with its long, thin, and largely naked abdomen. Staphylinids are commonly called rove beetles, although why I do not know, because they rove no farther than any other beetle. They look rather like earwigs, which, however, are placed in an entirely different order of insects, the Dermaptera. Both rove beetles and earwigs possess undersized wing covers (or elytra) that extend only about halfway down the animal's abdomen. In contrast, the elytra of most beetles conceal not only the insect's wings but also its entire abdomen. The typical rove beetle has a snaky, slinky look, and a fondness for narrow crevices. The complex maze of refuse in the kitchen compost is perfect for the species I am watching, which Dr. J. M. Campbell will later identify as *Ontholestes cingulatus*. It looks around alertly, after polishing off the fly, and then when I shift my feet, it scampers out of view into an opening between two decaying plums.

I remain as motionless as I can while scanning the surface of the mound. From tiny caves and cracks two, three, then four well-separated rove beetles creep out. They weave their way through their fantastic terrain, disappearing, reappearing, joining others similarly engaged. One medium-sized individual reaches some wet stalks of hay. It pauses and twists its abdomen around so as to probe the haystalks. Aha! Is this a female attempting to lay her eggs in the compost? My question is quickly answered when a wandering beetle comes around a melon rind and spots the presumptive female. The newcomer approaches with short, jerky steps and marches around to the rear of the other beetle, which warily withdraws its abdomen from a grass stem. The approaching male, for it is a male, uses the underside of his head to tap the abdomen of the female, for it is a female. She remains still and the male curls his abdomen down to contact his partner's genitalia on the tip of her abdomen. They copulate for a few seconds, and then separate.

The female walks a short distance forward and begins again to probe vegetable matter. To my surprise, however, the male does not resume his navigation of the compost heap, but instead follows right behind his recent partner. While she lays some eggs, he literally stands over her on stilted legs. She moves, he follows. They slip off into the obscurity of an opening in the compost, the male right behind his ex-mate.

Now here is a puzzle that convinces me straightaway to postpone my decision to transport the kitchen waste to the major compost heap. Instead, I secure a folding camp stool from the basement and return to set up an observation post near the kitchen compost in the shade of a twisted locust tree. The flies and bald-faced hornets buzz up and then settle down again as I balance myself on the camp stool.

The rove beetles that headed for cover both when I left and when I came back soon regain their equilibrium and step out from hiding places to carry on as before. I watch them to resolve the puzzle that has engaged my curiosity—namely, why should a male that has mated with a female remain with her instead of searching out new females to inseminate? Once a male has donated sperm to a female, there would seem to be no reproductive advantage to be derived from keeping her company, only time and energy lost to the hunt for additional partners carrying more eggs to fertilize.

I have a possible answer in mind, that the male is guarding his partner against other males, which would like to mate with her and so provide her with new sperm of their own. If these "additional" sperm entered the female, they might dilute or replace the first male's donation when the female fertilizes her eggs. (Fertilization occurs when a female releases eggs from her ovaries, after which they pass down the oviduct where they will encounter sperm, which have been set free from a sperm storage sac

*Rove beetle male
guarding mate*

attached to the oviduct.) If, by remaining near an egg-laying female, a male prevents her from acquiring additional sperm, then his sperm will be freed from competition with those of other males and so will have a better chance of fertilizing eggs that are about to be laid. Clearly, the guarding male's time spent with his recent partner may reap a return in egg fertilizations that more than compensates him for the loss of other sexual opportunities.

I know from other insect studies that mate guarding is associated with strong sperm competition between rival ejaculates. But what I don't know is whether this scenario applies to "my" rove beetle. I can find out whether it does fairly easily, however, provided I can observe the beetles doing certain things.

If the males remain with their partners to keep them from accepting other males' sperm, I should see some females mating with more than one male in sequence (if the females happened to be left unguarded). I should also observe "guarding" males successfully keeping rival males (and their sperm) away from their females. I am ready to watch. Will the beetles cooperate?

I have to be patient because my scattered subjects do not encounter one another with great frequency. Furthermore, the large majority of meetings either are between individuals of the same sex or else do not involve a receptive female. But I keep perfectly still on my camp stool and am eventually rewarded by observing a series of alliances between males and females, which mate and then stay together for some time.

By focusing single-mindedly on pairs and following them for as long as possible, I find males that follow their partners for up to fifteen minutes. During these consorts, unpaired males often find the squired female, at which point all hell breaks loose. The newcomer rushes toward the female; the "guarding" male always responds by interposing his body between that of the intruder and his mate, moving this way and that to keep his opponent at bay. Sometimes the ballet breaks down and the two males simply dash wildly in tight circles around the generally unperturbed female, which either lays eggs or walks slowly forward searching for good egg-laying sites. At other times, one male grabs the other in an unfriendly embrace and the two wrestle violently, belly to belly on the surface of the compost. The beetles' wild movements prevent me from seeing exactly what they are doing to each other, but I suspect that they are using their large, powerful jaws to add bite to their grappling.

By noting the relative sizes of the "guarding" and intruding males as a means of identifying them as individuals, I learn that males generally keep their opponents at bay. But at least three times, the original male was definitely chased off and replaced by a larger newcomer after a battle royale. The new male promptly courted and copulated with the willing female.

It sure looks as if egg-laying females remain receptive to ardent males, but I'd like more data. Therefore, I conduct an experiment using a long grass stem to poke and push following males, forcing them to abandon the females they are accompanying. In four cases, I manage to molest the male into leaving without also causing the female to sprint for cover. New males eventually find all four females, which promptly mate with these new partners. These observations delight me because they confirm that, yes, recently mated females do not lose their sexual willingness, so that it can pay a male to guard his partner. The "sexually promiscuous" females and violently "jealous" males behaved as I expected they would, if the mate-guarding hypothesis were true.

Having made some small sense out of nature, always a gratifying experience, I am free to resume compost building, which has been on hold for some time while I chronicled the sex life of *Ontholestes cingulatus*. I scoop up a quantity of moist kitchen waste from the mini mound and transfer it en masse to the much larger and still growing pile nearby. It twinges my conscience that compost mania has temporarily disrupted rove beetle life, but I console myself with the thought that I am merely providing them with a vastly greater arena for their sperm wars. Bald-faced hornets hum appreciatively as they circle the giant compost heap; committees of fruit flies reassemble on discarded plums; bacteria and fungi silently applaud my decision and diligently begin the work of making something out of nothing.

LAWN LOVERS

Who loves a garden still his Eden keeps, perennial pleasures plants, and
wholesome harvest reaps.

—Amos B. Alcott

Tablets

MY CURRENT ENTHUSIASM for vegetable gardening
is surely connected in some way to childhood
experiences, but the relationship is far from sim-
ple, as I shall explain. However, to the extent that
a fondness for gardening has some sort of genetic basis, I suppose I might
have inherited the trait from my paternal grandfather, who was a professional gardener on a large Cape Cod estate.

Gramps had his own magnificent vegetable garden at his home in Oster-
ville. As I was growing up, our family traveled to the Cape each year to
spend two weeks with my grandparents, a vacation I anticipated and
enjoyed, but most definitely not to garden. I was there to roam the beach,
to bird-watch for Hudsonian godwits and peregrine falcons on Monomoy
Island, and to gratefully consume my grandmother's amazing desserts,
which she produced for every evening meal. I spent very little time in my

grandfather's garden, and then generally only to pick blueberries for my grandmother's superb blueberry muffins and pies.

My father also had a garden, and during childhood I regularly spent time there. At the time, we lived in the country in southeastern Pennsylvania on seven acres of land that my father gradually converted into either lawn or garden, a huge mistake as far as I was concerned. The lawn required steady summer mowing and the gardens (there were several) also made insatiable demands on our time. My parents, my two siblings, and I became full-time mowers and weeders during the very season when freedom from school should have meant that I could spend every waking hour fishing the White Clay Creek that ran near our house through cow pastures and woodlots.

An unwilling member in the mowing brigade, I was also conscripted as a sprayer or duster in the battle against garden insects and fungi. One day my father assigned me to apply an antifungal concoction to his beloved Big Boy tomatoes. He gave me careful instructions about how to mix the stuff up, and I went about the job with scarcely a second thought. We were living in the pre–Rachel Carson era and so I regularly handled the wide range of toxins stored in our dank basement.

Although I have not yet detected any obvious damage from exposure to the insecticides, fungicides, and herbicides that I came in contact with in my youth, the Big Boy tomatoes came to grief from their encounter with me. Within a short time of dousing the plants with the spray I had prepared per my father's instructions, their leaves began to wrinkle and wither. By the time my father came home and went to admire his tomatoes, which had been on the verge of producing a bumper crop, he knew at once that something was amiss. With a little detective work, he figured out that I had poured the fungicide into the wrong sprayer, one that had recently held 2,4-5 T, a potent herbicide. Solving the mystery could not bring back a magnificent crop of tomatoes, but to his credit, he did not castigate me, instead suffering the loss with a fine resignation.

Nowadays my childhood gardening traumas seem so distant that I have no residual aversion of any sort to growing vegetables. In fact, I time visits to my parents' farm some sixty miles or so west of Washington, D.C., to coincide with their gardening season, which runs from April to September. In the summer in northern Virginia, the haze over the Big Cobbler and the more distant Blue Ridge blurs earth and sky on a humid summer day. On these days, a person listening to the call of a bobwhite quail from a distant hollow feels as if bird and listener were submerged in a diffuse sea, half water, half air.

Most of the farm's 300 acres are currently rented out to local farmers who grow soybeans or plant corn or run cattle, depending on the lessee of the year. I have no interest in industrial-strength agriculture, focusing instead on the much smaller vegetable gardens and lawns that surround the house, a small, two-story frame building that dates from the pre–Civil War era. My great-great-uncles sat on the front porch with their hound dogs in days long gone gazing out over cornfields and horse pastures. Part of what they saw in the nineteenth century has been converted into my parents' big rolling twentieth-century lawn. Whereas my Arizonan postage-stamp rye grass lawn in the backyard takes no more than ten minutes to mow by hand, my father spends a half day seated on an antique riding mower decapitating the grass every five days or so.

His green lawn flows down around a lane leading up to the house. It eddies out to encircle scattered walnuts and locust trees as well as two large rectangles of red Virginian clay, one to the north of the lane, the other well to the south, where my parents cultivate their vegetables. When rototilling the soil, my father regularly exhumes twisted horseshoes, fragments of stoneware, the occasional oyster shell. All attest to long occupation by farmers.

In the gardens, my father plants tomatoes, beans, peas, squash, eggplants, cucumbers, and corn, whose mature fruits he values in the extreme, perhaps even more than the Better Boy tomatoes that flourish on the farm. To harvest the ears of corn, however, he must compete with raccoon, deer, and corn borers (pestiferous moth caterpillars that burrow through rows of corn kernels). He sometimes wins this competition, sometimes loses.

During my week or two in Virginia, I relish the ample opportunities for yard and garden work available on 300 acres, a catalogue of possibilities that reveals how limited I am in Tempe, Arizona. What will it be? Pulling weeds out of the old graveyard in the remnant woodlot? Clearing honeysuckle from the rows of immature pines bordering the dirt road that leads to the house? Building or turning the compost heap by the lower garden? Gathering cowpies from the fields for the compost? Hoeing the flower borders? Picking stringbeans? Clucking over a zucchini fallen in battle against a ghastly, insidious, and remarkably fast-acting disease? Chopping poison ivy from a fence row? I select this last option only when I am feeling particularly masochistic, which is not often given the wide range of more acceptable activities.

Often, I give the entire lot a miss, opting instead for a round or two of insect watching. Today, the green june beetles are out and about, diverting me from the graveyard and the compost heap. Green june beetles are inch-long, chunky scarabs (that is, they belong to the family Scarabaeidae, a group held in special affection by the ancient Egyptians, although as we shall see, modern agriculturalists are a good deal less enthusiastic about certain crop-marauding scarabs). The species (*Cotinis nitida*) cruising low over my parents' spacious lawn is extremely common in the southeastern United States, no doubt in part because its immature stages feed on grass roots, a superabundant food in suburban and rural habitats alike. The scarab grubs nibble roots while tunneling slowly through the soil, becoming ever bigger in the process.

When the pale white larvae reach half the size of a man's thumb, their developmental program orders the flabby grubs to metamorphose into pupae. Pupae do not feed but instead remain still in their underground chambers until their tissues reorganize themselves once again, this time to form adults. After their external cuticle hardens, the now functional adults burrow up to the surface and ultimately into the light. Unfurling their folded hindwings from under their hard elytral covers, the mature scarabs take flight, zooming off into an environment totally new to them.

Today adult scarabs abound. They swirl over the lawn in great loops and swoops, some heading for blackberry patches where they bumble from berry to berry, cutting into the fruit, lapping up the sweet, nutritious juices while ruining or at least reducing berry quality from a human consumer's perspective. In addition, the adults gourmandize on various plant leaves, including those of grapes and many other fruits, which has made them unpopular with commercial agricultural interests.

But the beetles themselves often make the transition from consumers to consumed, thanks to certain insect eaters that find them highly edible morsels. I enjoy seeing blue jays convert green june beetles into blue jay chow, which is why I am watching birds and beetles instead of doing something more practical. With my back against a huge walnut tree, I scan downslope with binoculars for beetles and beetle eaters in the open lawn before me.

A blue jay perches low in a peach tree 200 feet away doing some scanning of his own. His head turns as he follows three beetles all flying in the same general direction. As the jay and I watch, one beetle turns off, circles in a tight loop, zigzags briefly into the wind, and then drops to the grass. No sooner is he down than the blue jay catapults from his perch and wings low over the ground toward the spot where the beetle plopped down. The bird lands and with a vigorous stabbing movement, pounds his beak into the earthbound beetle, then grabs the stunned insect and flips it into the air, before picking up the now disabled victim and thwacking it hard against the ground. The bird now adjusts the thoroughly tenderized scarab in his beak, giving it a couple of final pincer-like "bites," before raising his head to swallow the insect with an extravagant gulp.

After gobbling down his substantial snack, the jay flies back to the shade of the peach tree and resumes his lethal inspection of the lawn. He will make several more kills before losing interest in crushed june beetle. Over the years, I have also seen cardinals, red-bellied woodpeckers, and grackles tracking down prey and feasting upon them in much the same manner as this blue jay, although jays appear to be the most talented professionals when it comes to hunting lawn beetles.

Jays and other birds will on occasion attempt to pursue and capture flying beetles, rather than those that have landed. They succeed less than 50 percent of the time because, as you might imagine, airborne prey are

Blue jay flying toward a pair of scarab beetles

more agile and difficult to intercept than those blundering about on the ground, their legs entangled in grasses, their vision partly obscured. Walking beetles almost never escape when a jay or grackle tracks them down.

The obvious question—why do the beetles risk a walk across my parents' lawn, if the activity exposes them to vigilant predators as effective as blue jays? It could be that some of the beetles on the lawn are females searching for places to lay their eggs. In my experience, however, almost all beetles that run the predator gauntlet are males out hunting virgin females. When I play the blue jay, and run over to where a beetle has landed, I am apt to find him wedging through the mat of grass down toward a female nestled deep within the turf. Upon reaching his goal, the male mounts the female while everting his considerable aedeagus (that's a penis in entomologicalese), which curves underneath his body at an angle. With assistance from the probing male, this structure insinuates itself into an appropriate part of the female. Speed is of the essence here because several males may arrive almost simultaneously near the same receptive female, the various suitors drawn no doubt by an odor released by the sexually willing beetle.

I am confident that a sex pheromone is involved because males always fly into the wind before landing, often coming in with a bit of zigzagging, as is characteristic of insects tracking an odor plume. Moreover, when I capture some unmated females and place them in opaque vials with a wire mesh cap, male june beetles fly to the vials and land on or by them even though they cannot see what's inside.

Once a female has mated, she loses her sexual receptivity and almost certainly stops releasing sex pheromone. Thus, the premium on being the first male to reach a scent-producing virgin female. But if a male is to seize this glorious opportunity to pass on his genes, he must land near a prospective mate and make his way to her with all the risks that a stroll on the lawn entails. Males that refused to participate in the race to reach females would surely be less likely to end their lives as fractured bodies sliding down a blue jay's gullet. But they would also be much less likely to pass on the hereditary basis for their risk-avoiding behavior to offspring, because males of this sort would rarely, if ever, copulate.

Blue jays and cardinals in northern Virginia and elsewhere take advantage of the intense "damn the torpedoes, full speed ahead" sex drive of male green june beetles. Their quintessential male "attitude" is the evolutionary outcome of generations and generations in which male beetles that consistently came in second or third in the race to find virgin females left no descendants to carry on their second-rate or third-rate sex drive. Selection of this sort favors adult males that are blind to almost everything except the alluring scent of a virgin.

In this the green june beetle is far from unique. For example, consider the digger bee (*Centris pallida*), an insect I have studied at length in natural desert habitats in Arizona (because, alas, it has not taken up residence in my artificial desert yard). Although completely unrelated to the big green scarabs of Virginia, the bee's tactics for finding mates are remarkably similar. Male digger bees, like green june beetles, gather in numbers in places with relatively high densities of emerging females. The male bees, like their beetle counterparts, can smell emerging virgin females, which are sexually receptive. To reach potential mates, male digger bees land on the ground and actually dig down into the soil to meet females on their way out. Males that are exceptionally good at finding and digging up these hidden females have better reproductive prospects than those reluctant to get down and dirty.

But as is true for the beetles, males on the ground are at special risk to predators that concentrate on the vulnerable, distracted, less mobile individuals, the sex-crazed males that are on the verge of contacting a potential mate. Several species of thrashers as well as mockingbirds and cactus

Digger bee digging for female

wrens hang out in the areas where digger bees are looking for females. When they spot males on the ground, these birds run or fly to them, rendering their victims helpless with violent stabbing pecks or crushing bites.

Green june beetles and digger bees illustrate a general rule in animal behavior—males are far more likely than females to engage in mating activities that expose them to predators. Why should this be the case? One answer, widely accepted now, is based on the following question: What kind of male will leave the most descendants and so have the greatest effect on the evolution of his species? A moment's reflection reveals that a male's evolutionary impact will usually be a function of the number of sexual partners he has in his lifetime. The more females mated, the more eggs a male can expect to fertilize, and the more descendants he should produce that will carry his genetic lineage forward. In contrast, the number of offspring generated by a female will usually depend on such things as the number of eggs she can manufacture or the quality of her maternal care, and *not* on the number of male sexual partners that she tallies during her lifetime. Thus, in most species, today's males descended from males that spent their lives seeking as many mates as possible, whereas existing females descended from those that permitted males to expend energy and take risks in locating them, the better to devote their lives to securing materials for egg production or nest construction.

Because this pattern applies to both the green june beetle and the digger bee, males of the two species become vulnerable to predators through their single-minded pursuit of females, whereas females can afford to wait for mates, letting the males take desperate chances while searching them out. However, females can suffer too from male sexual efforts. Male digger bees and green june beetles may inadvertently expose a female's location to predatory birds that were keeping tabs on male movements. If so, a curve-billed thrasher or blue jay may secure a two-for-one deal, nabbing not only the male insect that attracted its attention in the first place, but the female that the male found shortly before his death.

I have even seen blue jays ignore a male beetle on the lawn in favor of whacking a female. In one instance, a jay picked up the male beetle, threw him to one side, and turned his cruel beak on the waiting female, which he killed and ate. I wonder if female green june beetles have a higher fat content than males, making them a more nutritious and desirable meal than males of their species. Whatever the reason, some male insects unwittingly betray the location of edible females to bird predators, making sex costly for females as well as for males.

ARIZONA SPORTS A surprising amount of carefully manicured grass available for insect consumers. Despite my excellent example, Bermuda, tiff, and Augustine grass still dominate Loyola Drive, being the choice of over 60 percent of the fifty-four houses on our street. Meanwhile, in the desert nearby, bursage, creosote bush, paloverde trees, and saguaro cacti are being leveled at a record pace to make room for more houses, more lawns, and more golf courses. For example, I note, from the newspaper, plans for two eighteen-hole golf courses near the Superstition Mountains to the east of Tempe. The golf courses, designed by Jack Nicklaus and his two sons, Jack II and Gary, will be the centerpiece for 1,900 upscale homes, townhomes, and villas, all in the "affordable luxury lifestyle" category. According to the *Arizona Republic*, "Nicklaus said each course will be different and special because of the beautiful land on which they are being developed." The manager for the project expects to attract people from towns to the east of Phoenix "who want to move even farther east now that growth has caught up to their housing communities."

Aside from persons seeking to escape "growth," perhaps the Arizonan relative of *Cotinis nitida*, a scarab called *Cotinis mutabilis*, will benefit from this lawn and golf course bonanza. The Arizonan green june beetle is reputed to feed in grassy compost heaps. I have occasionally seen it cruising about Loyola Drive on summer days, but not in any numbers. My gravel and shrubs offer it little encouragement. Still, we do have some other vegetarians living under our yard, as revealed by the presence of thin delicate tubes of clay surrounding some dried stems of a dormant globe mallow shrub. These tubes signal as clear as clear can be that we have termites in the front yard.

I discovered the termite tubes one late summer day after the start of the desert monsoon season. A rain had moistened the soil some days previously, permitting worker termites in their underground retreat to begin working the soil and digging tunnels to the surface. Foragers carried tiny pellets of moist soil in their mouths to upright globe mallow stems. Groups worked cooperatively to glue the pellets together in such a way as to form a loose-fitting tube around the dry stem. The tube shielded additional workers from ants and other enemies when they began to strip slivers of

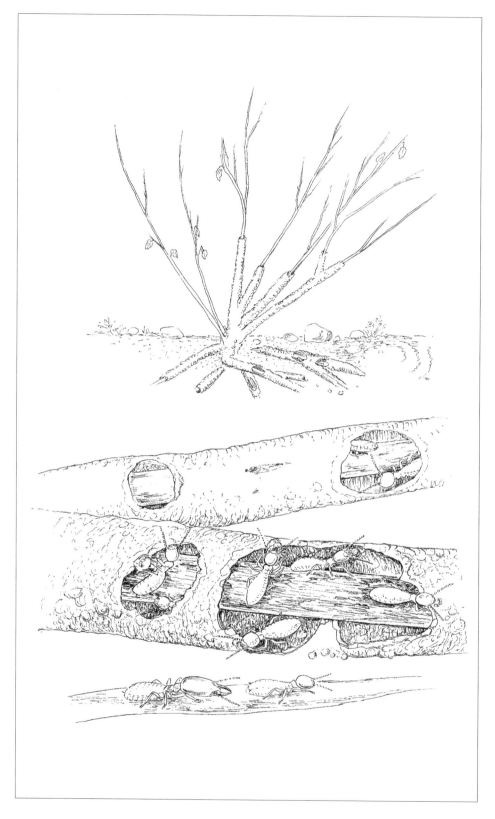

Termites working
on tubes on globe
mallow stems

tissue from the enclosed plant part. The tiny fragments of cellulose were carried back underground where they fed multitudes of other termites dependent on their fellow foragers for their meals.

A termite is one of those remarkably social insects whose ability to cooperate with one another is typically viewed with either admiration or fear, depending on whether one is an academic entomologist or a homeowner. Despite the fact that I belong to both camps, my predominant emotion falls heavily on the admiration side of the equation, especially because I believe that the little termites in the front yard belong to the native species *Gnathamitermes perplexus*, which is said to favor dead grasses, dried twigs and stems, and fallen cholla cactus bits, in which case they will probably leave the lumber in our house alone. Or so I hope.

Confident enough of our termites' harmless nature, I have encouraged them to stay around by serving up a number of large, dried cowpats, the crème de la crème of termite chow, as far as *Gnathamitermes* are concerned. Fortunately for them, if not for the rest of us, cow dung is a feature of almost every corner of the Sonoran Desert of Arizona (and Mexico), due to the affection that Arizonan (and Mexican) ranchers have for cows, huge numbers of which feed on private and public lands throughout the West.

In fact, my initial motivation for dropping cowpies onto the front yard had nothing to do with charity toward termites. Instead, I wanted a front yard with a true desert feel, and since cattle and their by-products are universal in the Sonoran Desert, I knew I had to import a few cowpies to attach the seal of authenticity to my creation. When I go out walking in the Superstitions or Mazatzals, I keep my eye open for just the right cowpats to bring home. A highly rated one must be dry and odorless but not so old as to be falling apart. It must also be attractively circular rather than asymmetric, and big, at least a foot in diameter and preferably closer to eighteen inches. Given the dominance of cows in the desert, I rarely have to hunt far for these treasures, which go into my daypack for later positioning by a paloverde or fairy duster in my yard.

My wife tells me that several neighbors have taken her aside to speak of the cowpies in our front yard. They claim to be unconcerned about their effect on local real estate values but rather wonder if I had a completely normal childhood. The local termites do not ask such hard questions. During fall and spring, when they forage on or near the surface, the termites immediately colonize my offerings by tunneling into the cellulose pie.

There in total darkness, they excavate bits of food to carry down into their subterranean home for nestmates. When I carefully pick up my cowpat, I can usually find several pale white workers clinging to the underside of the pat, their work on behalf of the colony interrupted for a moment. According to the termitologist William L. Nutting, a typical colony of *Gnathamitermes perplexus* consists of about 10,000 individuals, so that I am seeing only the top of the tip of the iceberg when I count the few worker termites clinging to a cowpie.

Where my front-yard *Gnathamitermes* came from in the first place mystifies me, but perhaps they are progeny of colonies that occupied nearby back alleys or possibly even a neighbor's house. In any event, my colony or colonies almost certainly formed on a late summer day when there was a nuptial flight on Loyola Drive. The nuptial flights of termites begin when sterile worker termites usher their reproductively competent brothers and sisters to the surface. The sterile workforce, having reared their siblings on behalf of their parents, a resident king and queen, now sends them out to be fruitful and multiply. The reproductives take wing, fluttering up to join many others of their kind simultaneously released from other colonies.

After a short flight, no more than a hundred meters, the termites rain down on the ground, if they have survived running a gauntlet of foraging birds and hunting dragonflies. Once earthbound, the alates flick off their wings, breaking them along a fracture plane designed for the purpose. Females then adopt a peculiarly rigid stance, body motionless, abdomen raised high in the air. While in this position, they probably release a sex pheromone, which draws males to them. Once a male has tracked a potential partner down, he strokes her body with his antennae and mouthparts. In response, the female lowers her abdomen and turns to lead her boyfriend away. The now wingless male runs after his equally wingless mate, keeping in contact by touching the tip of her abdomen, as she searches urgently for a small crevice in the ground. Predatory ants may terminate this phase of the union unless the pair are fortunate enough to find what they are looking for quickly.

If they are lucky, the pair will survive to dig out a small chamber in the soil, which termite experts have charmingly labeled the "copularium." There they will indeed copulate. Their first offspring will be their helpers, sterile assistants who build enclosed foraging trails and snip fragments of cellulose within the darkness of protective mud tubes. The worker offspring repel ants and other unwelcome intruders and help care for the eggs laid by the queen and fertilized by the king. They are neutered extensions of their parents, and there may be hundreds of thousands, even millions, in mature colonies of certain species. The nests of many Australian and African species balloon to monumental proportions. Were our local desert termites to produce a massive clay "watchtower" about twenty feet high and eight feet in diameter at the base, I might be considerably more alarmed to discover an incipient termite colony in the front yard.

Of course, if termite kings and queens produced *only* sterile offspring, their reign on this planet would be short indeed. They avoid extinction in a single generation by eventually producing some offspring with reproductive potential, although this may take up to ten years in some cases and will require assistance from vast hordes of sterile worker progeny. These workers actually rear future royal couples to adulthood and determine when to let them out for their nuptial mixers, which occur on just a few days each year for most species. The king and queen whose workers now roam through imported cow dung on my front yard probably came from

Giant termite nest

two different colonies. Luckily they found and also acquired a congenial chunk of real estate, complete with a cowpie or two to help sustain their intricate and complex society in the years ahead.

Although I rarely see more than a couple of termites at a time, and then only when I inspect one of the yard's cowpies, it's enough to know that they are part of my world, invisible neighbors, but neighbors nonetheless. And the complexly social termites are just one of a vast armada of insects whose lives go by largely underground and out of view. Every time I walk on my front yard, I know that whole worlds thrive beneath my feet, but not beneath my interest.

NATIVE STINGERS

A few years ago a gentleman came up to me when I was mounting wasps at a picnic table in a Missouri state park. "What is the purpose of a wasp?" he asked. Had I been a lepidopterist, he doubtless would have asked the purpose of a butterfly, though I am not sure what he would have asked had I been an anthropologist.

—Howard E. Evans
Life on a Little-Known Planet

THOSE SMALL CREATURES able to resist us with a sting suffer greatly from human intolerance and the widespread inability to see the beauty in all living things. Take scorpions, for example. I clobbered a scorpion to death once, fearing its sting held high at the tip of its curled "tail" and loathing its Grade B horror film appearance. I have since developed, if not a fondness, at least a modicum of respect for scorpions. But my current attitude is shared by few others. As a result, no laws protect the scorpion, no organization devotes its energies to the Prevention of Cruelty to Scorpions. On the contrary, a company called Western Heritage goes out of its way to concoct a ghastly death for tens of thousands of scorpions annually. Western Heritage is a Scottsdale wholesale company that specializes in scorpion paperweights, bola ties, and refrigerator magnets. For $12.95 you can buy a lump of plastic resin that encases an Arizona scorpion or perhaps you would prefer a plastic mausoleum containing a black widow spider or a tarantula. They are available in souvenir shops at Sky Harbor International Airport in Phoenix.

To prepare a souvenir of the Wild West, Western Heritage pays pickers to wander around in the desert at night with a black light. Under ultraviolet radiation from the light, a scorpion literally glows, permitting its easy detection and capture. The picker's catch is then transported to pourers who decant liquid resin into molds and then stuff living scorpions into the plastic after it is partly hardened. According to the *Arizona Republic*, Doug and Linda Parker performed this macabre operation about 120,000 times in 1994. Apparently, it beats farming.

A few scorpions pack a sting powerful enough to do real damage, but the "bite" of most species is not even remotely life-threatening and no more painful than the sting of a bee. Of course, most people do not care to be stung by a bee either. As an entomologist devoted to the bees and wasps native to the southwestern deserts, I have been zapped more than the usual number of times by a greater number of species than the typical person. Although I generally study male wasps and bees, which, happily, lack stingers, occasionally I err in the sexual identification of a netted specimen, and I am the first to know about it, since female wasps and bees do not hold back when being extracted by hand from an insect net.

Fortunately, the sting of most desert bees and wasps carries only a sharp but transient jolt of pain, with little or no subsequent swelling or discomfort. If one has to be stung, I recommend one of these species, rather than a honeybee, which is not only more potent but an "unnatural" introduced species to boot.

All rules, however, come with their exceptions and here I speak of any of the larger paper wasps of the genus *Polistes*, a group very much native to the Americas (there are seventeen species or thereabout in North America alone). The pain of the sting of *Polistes annularis* has been ranked at 3 on a 0 to 4 scale by Justin Schmidt, who interviewed numerous individuals unfortunate enough to have been assaulted by a diversity of bees, ants, and wasps. Honeybees check in at level 2, and only a select few bees, ants, and wasps have been awarded level 3, with fewer still achieving the dreaded level 4. Another of the level 3 species is a mutillid wasp with the common name "cow killer," which may give you some idea of what it feels like to be nailed by a level 3 stinger.

Paper wasps get their chance to do a number on people because they like to build gray papery nests around the eaves of human habitations and outbuildings. An abundant species in Arizona, *P. annularis* belongs to this group of "civilized" paper wasps that locate their hanging nests on buildings where they are sheltered from sun, wind, and rain. The wasp also nests in the dense shrubbery of the sort nurtured by homeowners in Phoenix and it visits swimming pools for water.

I can recall almost every instance when I inadvertently placed body part next to paper wasp stinger, but my meeting with *P. annularis* on my home ground made a stronger impression than most. I had just begun to pull down my old toolshed in the backyard, anticipating its replacement with a new, sturdy, prefabricated one, a sort of small barn with a green roof, which I thought would bolster my image as a gardener. The old shed leaked from every metal seam, especially from its decaying, swaybacked roof. I had covered part of the roof with a large board in an unsuccessful attempt to keep the shed's interior dry. The board had to be removed if I was to tear down the rest of the decrepit shed. Not a problem, or so I thought, when I began work without seeing a small colony of *P. annularis* in the space between the board and the roof.

As I lifted the board, I glimpsed the nest and its retinue of wasps and almost immediately I felt three red-hot needles inserted into my face with exceptional vigor. Defense of nest is top priority for most paper wasps,

Polistes *paper wasps on nest*

which have to deal with a host of birds and small mammals that would remove the plump juicy larvae occupying cells in the paper nest, unless powerfully dissuaded. I instantly lost all interest in shed deconstruction. Powerfully dissuaded, I dropped the board and retreated at full speed, various wasps in hot pursuit. I burst into the kitchen, pushing the door shut behind me to keep my sadistic assailants away, and turned to my wife and children to tell them my story and receive their affection and concern. Instead I was greeted with nervous laughter, my mouth and cheeks having already swollen into a grotesquerie that my family found amusing, although they subsequently assisted me in preparing an ice pack. When I checked in a mirror, I saw a complete stranger, someone wearing one of those misshapen ceremonial masks that the Iroquois once constructed. The big-lipped asymmetry of my face was no cause of hilarity for me, since I wondered if the disfigurement might not be permanent. It was not, although the swelling and pain stayed with me long enough to teach me a lesson that I had no desire to learn.

Despite the ferocity of their nest defense, paper wasps are exceptional insects well worth having on one's property, provided one knows where they are. If you approach a nest slowly and respectfully, you can watch the wasps at work with little risk of being attacked. And if you get to know the colony well, you may even figure out when the wasps are getting a tad grumpy, which they signal by raising their wings and bending their abdomen to the side to show off their fearsome stinger. These threats generally precede outright assaults and so are worth noting.

A patient and careful paper wasp watcher will witness many entertaining and instructive wasp activities, such as the distribution of food among colony members. When a returning forager wings in with a well-chewed caterpillar in its jaws, the predator may transfer part of the packet to other adults on the nest. If there are young to feed, the forager and its nestmates will soon be seen poking their heads into the brood cells where translucent larvae wait for their mouth-to-mouth meals.

Many other elements of paper wasp social life have been worked out by researchers willing to spend hours perched on ladders recording who does what to whom in colonies whose members have all been captured, anesthetized, and given distinctive marks or numbered tags. These researchers have documented several different kinds of female coalitions living on paper wasp nests. For example, the memorable cluster of avenging females on the nest of *P. annularis* that I disturbed may have been completely unrelated individuals that joined together to build the nest, feed the larvae, and protect the lot against man and animal alike. Or, the nest might have been founded by a group of sisters who were able to recognize each other even though they had spent the winter apart, hibernating in various palm trees in the neighborhood. Or, the nest might have been occupied by yet another kind of family, one composed of a mother and her daughters.

Paper wasps exhibit the kind of flexibility in social arrangements that we normally associate with certain supposedly "higher" mammals or birds. Figuring out why the various kinds of female groups assemble in *Polistes* requires that we identify the advantages and disadvantages to the participants. Some of the benefits seem obvious enough. First and foremost, a mob of stinging females surely is a more powerful deterrent to a hungry predator than a single female. In addition, coalitions can better deal with enemies of a different sort, namely nest usurpers from their own species. Nests with several occupants are always protected by at least one attendant. In contrast, a nest with a single foundress will often be unguarded, when, for example, the solitary female leaves to forage for herself or her brood. The temporarily abandoned nest is more susceptible to raiding females than one with a caretaker or two. The interloper may be able to prevent the returning owner from regaining control of her home. The successful nest thief saves time and energy—no need to build on her own—and she will either eat the larvae in the stolen nest or enslave them when they become new adults.

Given the defensive edge for communal nesters, it seems surprising that there are ever any nests with a single female resident, and yet there are, particularly at the start of nesting seasons. We can deal with this puzzle by recognizing that the benefits of living together are not shared equally among nest occupants. Typically, a single dominatrix suppresses the reproduction of her companions, who in effect work for her and her brood. This top female lays all (or almost all) the eggs that are so carefully tended and fed by her nestmates. Should an uppity fellow female sneak an egg of her own into an empty cell, the head female usually sniffs out the egg and eats it, emphatically putting an end to her rival's reproductive ambitions. If you have the courage and inclination to watch a small but growing nest, you'll very likely see a female poking her head into one cell after another, antennae twitching as she checks out the odor and thus the provenance of the elongate white eggs in those cells.

With the reality of one-female rule in colonies of paper wasps, the puzzle gets reversed. Why would any female choose to join a colony in which she will be exploited by the dominant queen? Paper waspologists have decided, after a host of behavioral studies, that different kinds of females have different things to gain or lose by being part of a society of wasps. First, a subordinate female may only be temporarily prevented from reproducing. If the dominatrix is injured when fighting with a nestmate or intruder, or if she is disabled by a raiding blue jay, one of her long-suffering colleagues may then have her day in the sun. It is true that the probability of inheriting a colony may be low for one of the currently non-reproducing females, but being part of a group could still provide on average a higher reproductive payoff, to use the economic jargon beloved of behavioral scientists, than the corresponding payoff to subordinate females that try to make it on their own without any helpers.

Second, subordinates who are related to the dictatorette stand to gain something even if they never ascend to the queenship of their colony. If their activities effectively boost the reproductive output of a sister, then the helper gains genetic representation in the next generation indirectly by virtue of having made survival possible for some extra nieces or nephews. These relatives will carry forward some of the helper's genes since she shares a recent common ancestor with her nieces and nephews.

Finally, consider the case in which daughters become drudges working on their mother's nest. They too can potentially perpetuate their genes by entering the wasp nunnery and spending their celibate days helping a domineering mother make many more sisters than their mother could have produced by herself. If some of these "extra" sisters are the dominant reproducing kind, then their self-sacrificing relatives can pass on some of their genome, the part that they share with those sisters who go on to have larvae of their own.

The story may be more Machiavellian if mothers ensure that some of their daughters have relatively low body weights and thus little chance to dominate rivals, since weight is a key factor in determining who wins contests for control of nests. In which case, lightweight daughters have little to gain by leaving home to find reproductive happiness elsewhere. Instead their only profitable option is to remain docile workers at their mother's beck and call.

Checking all these possibilities requires information of all sorts. Are assemblies mother-daughter affairs, and, if not, just who are the colony-mates? In addition, it would be good to know the average number of descendants produced in associations of different compositions and sizes, the frequency of turnover in nest "ownership," and so on. No wonder a legion of researchers plugs away on paper wasps, checking DNA fingerprints or their equivalents, marking females ever so gingerly, recording the survival chances of solitary and social nests, and the like. Thanks to the enthusiasm of biologists willing to get close to fully armed paper wasp females, we now know much about the social strategies of these insects, which assist individuals in propelling their genes into future generations of intimidating stingers.

IF YOU INSIST, you can place a dollar value on having paper wasps around the house and garden. As I mentioned, foraging paper wasps are predators with a sweet tooth for moth caterpillars. So if your garden is afflicted with cabbage loopers or something similar, then you have a nonacademic reason for tolerating paper wasps in the vicinity. Some commercial agri-operators have gone so far as to put up small wooden shelters in their fields in the hope that paper wasps will take up the offer of free housing and then repay the farmer by killing crop-damaging pests.

It is, however, easier to justify bees when it comes to the typical garden, which almost invariably has some zucchini squash. Zucchinis are to a summer garden in Tempe what Swiss chard is to the winter garden, a reliable, steady overachiever guaranteed to generate more vegetables than any ten vegetarians could consume. A cheerful, almost exuberant plant, with expansive green leaves sheltering brilliant orange-yellow flowers, zucchinis announce for all the world to see, "Here is a gardener!"

In my eagerness to get the squash harvest under way in the early spring, I get directly involved in zucchini reproduction. First, I check to see whether the plants have produced both male and female flowers—zucchini is one of those plants that has separate-sex flowers. (Real botanists tend to get upset at anyone who speaks of male and female flowers, but their reasons are too semantic and technical to go into here, and so I will persist with these terms at the risk of irritating a botanical reader or two.)

If my zucchinis have generated the two types of flowers, I pluck a pollen-producing "male," one whose petals emerge straight from a thin hollow petiole. Carrying the flower with its conical central stalk laden with sticky zucchini pollen, I next find a suitable "female" flower, one whose flower petals emerge from a green squashlet. I press the male flower against the convoluted carpel of the female flower, which rises like a fist from within a cup formed of radiant petals. Pollen grains galore stick to the top of the "fist," thanks to the flower union that I have arranged in the manner of a marriage broker.

One of my transferred pollen grains will in short order send down a tube through the tissues it rests upon, snaking its way toward an ovule at the carpel's base where fertilization may occur. If this blessed event takes place, the squash fruit will receive orders from the plant to grow and develop into a harvestable vegetable (eight to ten inches long, but no more, as far as I am concerned), one worthy of transport to the kitchen. If successful pollination does not occur, however, the fruit will not develop and soon turns into a yellow, withered reminder of a missed opportunity.

Why a squash plant produces separate male and female flowers is an intriguing issue. Many hermaphroditic plant species make flowers that

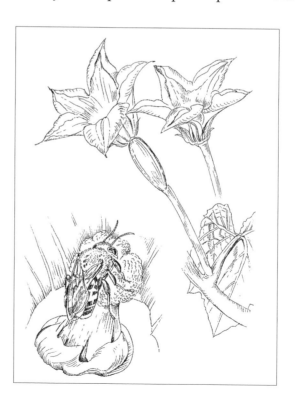

*The hairy squash
bee in a zucchini
flower*

contain both pollen-producing stamens and ovule-producing carpels rising side by side within a shared cloak of petals. These flowers can generate pollen for export while also receiving pollen in order to form seeds surrounded by a fruit. Given the abundance of double-function flowers, why do squash insist on making one-sex flowers?

One possibility is that, sometimes, if a squash plant made dual-function flowers, it would get no return for its investment in the large and expensive female part with its attached immature fruit. Such a structure requires energy expenditures that, for example, a small plant with few leaves and little access to water might fail to muster. The disadvantaged plant, however, might still be able to manufacture some relatively cheap pollen for transport to other individuals, thereby, maybe, leaving descendants via the male route without having to waste energy in making the female components of a flower.

Whatever the evolutionary reason for separate male and female flowers in squash, I soon stop moving pollen about as the spring progresses. Ending my role as fertility counselor for zucchinis, I let nature take its course. The change in my behavior correlates with the number of uneaten zucchinis in the refrigerator and with the growing abundance of another pollinator, a bee that gladly does my work for me. The bee in question is called *Peponapis pruinosa* (the hairy squash bee, as translated from the Latin). It is a squash pollinator par excellence that visits only cultivated squash and closely related wild gourds to satisfy its need for pollen and nectar. Squash nectar sustains adults of both sexes. Squash pollen is gathered up by females and carried to underground brood cells. There the female bees mold it into a ball before laying an egg on the provision mass. The larva that hatches from the egg will feast on the zucchini "bread" left for it by its mother, and eventually become an adult in the next generation of squash-dependent bees.

The importance of squash pollen and nectar for the hairy squash bee suggests that they should be good at getting these resources in competition with other potential consumers, and they are. I tend to wake up around daybreak in the summer, and often make a dawn pilgrimage to the garden to admire and encourage my plants. Even at this early hour, whenever I inspect a zucchini flower I am likely to find a hairy, rufous bee at work. In fact, students of *Peponapis pruinosa* in California have documented that the bee generally begins foraging thirty to sixty minutes before sunrise! At this time the bees may well have to use the odor of squash plants to get close enough to spot flowers in the pre-dawn.

The determination to get at squash flowers first thing in the day makes sense to students of zucchini who know that the squash opens new flowers in the very early morning, making it possible for the early bee to reap a fresh bonanza of pollen and nectar. No wonder that the hairy squash bee is a superior pollinator of zucchinis, better than the introduced honeybee, which will forage at these flowers, but not with the same early-morning dedication of *Peponapis pruinosa*.

Why does a wild North American bee have such fondness for an Italian squash? The answer to this question carries us back to zucchini beginnings. Domesticated summer squash—zucchini are merely one variety—belong to the genus *Cucurbita* with its approximately twenty-seven wild and five cultivated members. The summer squash have been assigned to *C. pepo* ("pepo" is from the Greek "pepon," meaning mellow or ripe, whereas "squash" is from the word "asku'tasquash" used by a Massachusetts Indian tribe to mean "eaten while green").

Most of the cultivated species of *Cucurbita* now have a worldwide distribution. In contrast, the wild members of the genus occur in Latin America with a center of diversity in Mexico. Therefore, *C. pepo* probably traces to a Mexican ancestor with domestication getting under way perhaps as much as 10,000 years ago, or shortly after the spread of people through the New World. Since all of the wild species have bitter flesh, the first contacts between *Cucurbita* and people probably arose when someone discovered that the seeds within the bitter fruits were actually quite tasty, especially when lightly roasted.

I like the speculation that squash "agriculture" originated in the serendipitous growth of the plants from trash heaps near the temporary camps of hunter-gatherers. Once people saw how easy it was to grow squash in unplanned compost heaps, they might have decided to help matters along by throwing a few seeds into discard dumps, coming back later to harvest their squash and remove their seeds. Mutants that happened to produce squash with edible flesh might then have been discovered and cared for, setting the stage for additional selection by early agriculturalists. These processes have changed cultivated squash considerably, so much so that its closest living relative is not obvious, although some likely candidates have been identified.

In any case, the mother of all zucchinis was almost certainly a Mexican plant that attracted native pollinating bees in the genus *Peponapis*, probably *P. pruinosa*, which still occurs in Mexico. But the hairy squash bee has spread greatly with the distribution of cultivated squashes throughout North America. Indeed, purposeful attempts have been made to introduce *P. pruinosa* to squash fields in Hawaii and even New Zealand. It is one of those relatively few innocuous species that has benefited greatly from the activities of man, thanks to its relationship with a plant of value to us.

IN ARIZONA, AS elsewhere, spring is followed by summer, which leads to the decline of my zucchini plants and the eventual disappearance of the hairy squash bee from my garden. Arizonan summers are quantitatively, if not qualitatively, different from the tamer versions experienced in other parts of the United States. The intense searing heat begins in late spring and lingers for months. In anticipation of such unending summer, many of the plants in our front yard wrap things up long before the really hot weather arrives.

The warm days of midspring suffice to cause the flowers of poppies and bluebells to lose their petals, provided that their native bee pollinators (which do not include *P. pruinosa*) have done their job. The ovaries at the base of the flower expand, green seeds growing within supportive tissues. Poppy seeds soon turn black inside graying fruits. As the plant's foliage withers, the fruits become small explosive devices, poised to split open when rattled by the wind or brushed by a mourning dove. With the violent rupture of a pod, seeds go flying. The plumper fruits of bluebells also dry out and eventually open, sedately spilling the mature, reddish brown seeds on the gravel for months of quiet dormancy—if the mourning doves and house finches do not get them first. Sometime toward the end of spring, I get a rake and start sweeping up the vegetative debris resulting from the decline and fall of the annuals, once green and fleshy, now brown and trashy, an eyesore for the neighbors.

Brittlebush is a perennial and therefore my front-yard brittlebushes will live to see another spring, except for a plant or two that I remove on the grounds that we now have too many of these exuberant propagators. However, as blast furnace June rolls in, the brittlebush seem to hunker down like a trapped firefighter under his last-ditch aluminum survival blanket. Their summer leaves turn white with a mat of hairs that greatly reduces their efficiency as photosynthetic sugar producers, a price paid for the improved heat-reflecting properties of these leaves, which keep the shrub barely alive during the dog days ahead.

The flower stalks of brittlebush, once green and supple, now become rigid, straw-colored stems with a tip slightly thickened and curled, like a tiny clenched fist. The dead stems rise above the foliage. The dried rods will eventually crumble and break, but for the moment they provide mildly defiant testimony of the plant's recent burst of reproduction.

In the early evening of July 18, I attempt a stroll through the front yard, as the superheated day starts to cool, which makes it possible to get out of the house for a bit. I carefully breathe in the still painfully hot air, noting with satisfaction the end-of-the-day shadows stretching out from brittle-bush, paloverde, and ironwood.

A flurry of movements in and around the dried flower stalks of the big brittlebush in the middle of the yard catch my eye, and I wander over to see what's up. There I find a mob of small native bees flying from flower stalk to flower stalk. Individuals drop out of the "swarm" to settle on the outer parts of stems, often where other bees have already perched, their bodies pressed close to stems that they grasp with their jaws. Newcomers bump earlier arrivals; previously settled bees react by kicking back or shifting position or flying up to join the others cruising about in search of a landing strip. As I come closer, my movements alarm some perched bees, which take off only to resettle on stems here and there in the shrub.

Mourning dove on gravel

On one stalk, six bees line up, each member of the group practically touching its neighbor. Eleven bees occupy another stalk, one after another like planes waiting for takeoff on a runway. Here is a singleton on a solitary perch; two additional bees have found another thin stalk that they alone share.

I am delighted because what I have here is a sleeping cluster of male bees, which I later learn from the bee biologist Wallace LaBerge belong to the species *Idiomelissodes duplocincta*. Males of these bees can be identified by their unusually long and supple antennae. When I come out again in the dusk, all the bees have settled down and rest immobile in the growing darkness. They will spend the night together.

I am up early in the morning of July 19 to count the sleeping bees, exactly fifty males in all, twenty-seven of them sleeping on a single stem near the center of the brittlebush. I wait for the bees to get up, which they eventually do, although they seem in no rush. They do not even stir until the sun climbs well above the eastern horizon and temperatures are rising too. Finally, the males leave their perches one by one, darting around the bush, zipping off. I hunt for them in the rest of the yard, but they are gone, vanished to wherever they spend the greater portion of the daylight hours. Will they come back in the late afternoon?

They will indeed. Or at least males of the same species assemble for the ritual of finding a spot on which to lay their head for the night to come. As on the previous day, there is much toing and froing, perch shifting and jockeying for position, before the males get themselves sorted out and lined up, usually with fellow males in close proximity. Furthermore, it seems to me that many of the bees are clustered on the same prominent stem that was so popular the preceding evening, even though there are dozens of vacant dried flower stalks to choose among.

I feel privileged to have acquired a sleeping bee aggregation in the front yard. It is an odd phenomenon that cries out for investigation, and what closer laboratory for behavioral research than my own front yard? I have three questions I'd like answered. First, why are the sleeping bees a males-only club? Second, do the same males return to the same brittlebush, perhaps the same dead flower stalks, from night to night? Third, the $64 question, what benefit, if any, do they derive from sleeping together?

Let's examine these puzzles, one by one. The question of the exclusively male sleeping club takes me first to the library to hunt for published infor-

mation on the genus *Idiomelissodes*. My search yields a single paper in the *Pan-Pacific Entomologist* written in 1975 by a T. J. Zavortink, who reported that females collect pollen from barrel cacti, an interesting observation but not entirely pertinent to my quest. The lack of attention given these bees by my fellow entomologists probably stems from three factors: their geographic range (the desert Southwest), their flight season (summer), and their daily activity pattern (middle of the day). This combination spells H-E-A-T in a big way. Even avid entomologists sensibly prefer more temperate conditions in which to pursue their profession.

Fortunately, bug watchers working in many places and seasons have over the decades accumulated a good deal of information on bee species closely related to *Idiomelissodes duplocincta*. The standard features of these relatives probably apply to my front-yard bee as well, in which case female *Idiomelissodes* excavate vertical or sloping shafts into the soil that could be a foot or so deep. Brood cells probably come off the lower part of the main tunnel. The nest owner fills these side chambers with food for her offspring, probably using the pollen of barrel cacti, judging from Zavortink's report of the pollen preferences of my bee. On each mass of pollen the female lays one egg before sealing off the cell and starting over again on a new brood cell. The egg hatches into a grub that gobbles up the food so carefully provided for it by its mother, before eventually metamorphosing into an adult bee that digs its way to the surface to participate in a new round of adult sex and nesting and sleeping.

Females in this ground-nesting coterie of bees usually spend the night in their underground burrows, where they are protected from the elements and out of view of potential predators. Therefore I suspect that female *Idiomelissodes* sleep inside their nests as well. But since males of this and other ground-nesting species do not help their mates build their underground tunnels, they do not have burrows in which to retreat at night. Having spent their daylight hours searching for mates in order to leave as many descendants as possible, they are in a bit of a bind when dusk comes and females have hidden themselves away. Where can they go?

Although I found some reports of males in species related to *Idiomelissodes* dozing in abandoned burrows or other crevices in soil, many male bees curl up in flowers or snooze the night away while hanging on to stems and twigs of shrubs and trees. In quite a few species, males seek out the company of their fellows when sleeping *al fresco*. The communal sleeping habits of certain native bees make them conspicuous, which may explain why scientific papers on the phenomenon began appearing almost a century ago.

However, most modern investigations of sleeping bees have been a bit perfunctory. Typically, entomologists have limited themselves to comments on the degree of aggregation of sleeping males and their choice of a nighttime roost. I will therefore have to find out for myself whether the same males of my bee return to the same stems night after night. As I march out to the front yard in the morning of July 20 full of investigative enthusiasm, I come equipped with some small vials of acrylic paints and a thin grass stem with which to apply droplets of paint to the backs of the torpid males. At this time of day, they are so sluggish (sleepy?) that I can mark them without causing undue alarm to either the bee selected for decoration or his neighbors. In no time at all, I have given ten males in one large cluster red marks, and then it's on to another group to mark six bees with white dots.

The following day (July 21), among the sixty-nine males assembled at the brittlebush I find eight "reds" and three "whites," a return rate of nearly 70 percent. But only three sleeping males are perched on the same dead brittlebush stalks that they occupied the night before. Although over the short term they are faithful to a particular sleeping bush, their choice of sleeping stems and sleeping companions within the bush can change from evening to evening.

Marked males continue to show up over the next several days, with four males bearing paint dots appearing as late as July 24. Clearly, some male bees are creatures of habit, with a strong preference for a traditional sleeping bush after a day spent afield.

I am intrigued by the tendency of those males that do assemble in the evening to bunch up on a single stem. Remember that on my first day of investigation into sleeping bees I found more than half of all those present in the brittlebush packed onto a single dried upright stem among many seemingly identical stems. This pattern persisted in the days that followed; 43 percent of seventy-two males were on one stem on the evening of July 19, and 29 percent of ninety-seven males were lined up on this most popular stem on the evening of July 22.

What cues cause my male *Idiomelissodes* to form their sleeping societies? One possibility is that the bees prefer a particular stalk based on its distinctive odor or shape. Another possibility, however, is that the bees like certain positions within the sleeping shrub, and have no special attachment for one stem as opposed to another, as long as it is appropriately situated. To discriminate between these possibilities, I wait until the bees have departed on the morning of the 23rd and then cut off the top 10 inches of the current favorite perch stem at what we will call site A in the shrub and replace it with a currently ignored stem of equal length taken from elsewhere in the shrub at what we will call site B. The replacement stem is splinted in place at site A with masking tape wrapped around its base and the stump of the original stalk. In the same way, the old favorite gets transplanted and taped to the stump of the replacement's stem at site B.

That evening I wait with what is generally said to be keen anticipation for the return of the sleepy bees. They show up on schedule, and begin to zip and swirl about the brittlebush stalks, landing, flying up again, trying a new spot, and finally settling on their beds for another evening. With the bees finally snuggled down, I count where they are: thirty-one clutch the old favorite stem that I moved to site B, zero occupy the replacement stem at the spot (site A) where the old favorite once stood.

I consider the results wonderfully conclusive. The males apparently are not selecting a sleeping site on the basis of its position within the brittlebush shrub, but rather have a preference for a specific stem. The bodies, particularly the jaws, of many bees are highly scented and Zavortink reported that *Idiomelissodes* have a characteristic odor as well, considered either pleasant or unpleasant, depending on the intensity of scent. When I sniff a group of bees in the early morning, after first checking up and down the street to make sure that no one will catch me in what my neighbors might consider a bizarre activity, I find that, yes, there is a special odor to the bees. I suspect that this bee scent, which I find neither disgusting nor ambrosiacal, enables males to recognize a stem used by their kind on previous nights. Lingering bee aroma could help them satisfy a preference to sleep in a used bed.

Although the results of my first experiment were highly clear-cut, always a pleasure and relief for an experimenter, I decide to repeat the experiment. On the 24th of July, I do another switch, moving the old favorite from site B to a new location (site C) several feet away on the outer edge of the brittlebush, and taking the stem that it supplanted back to be its replacement at site B. As so often happens in science, the results this time contradict those of the initial experiment. My first check in the evening of the 24th reveals not a single sleeping male on the old favorite stem at site C, nor are males settling on its replacement back in the center of the shrub at site B. Instead, small clusters of males occupy a considerable number of stems *near* site B, where the old favorite stem had been transplanted in the first experiment.

I decide to postpone my count until the next morning, when it is easy to approach the torpid sleepyheads without disturbing them, the better to conduct a bee census. Alas, when I come out just after dawn, I find that something has disturbed the males and caused most to disperse already. However, although thirteen stems are still occupied by one or more sleeping males, what had once been the favorite stem is not taken by a single individual.

Hmmm.

Clearly, bee scents alone do *not* determine where males choose to sleep. Perhaps position in the shrub does play a role after all, with the bees rejecting peripheral locations in favor of stems near the center of the shrub. This possibility deserves to be checked.

So I get out the scissors, snip the masking tape once again, and organize yet another "bed" shift. This time I take the old favorite stem back to site A, the spot where it had come from in the first place, removing the stem that had replaced it during that first experimental shift. In the evening of the 25th, ninety-three males are kind enough to show up for a good night's sleep in the brittlebush. None, not one, settles on the old favorite at site A. All are gathered together on stems in the vicinity of site B instead.

Hmmmm.

Okay, I think I have it. Since the old favorite stem had not been used by sleeping males on the 24th, when I put it in an unacceptable outer location in the shrub, perhaps any male scents had largely evaporated from it by the evening of the 25th. By then, even though the old favorite was back in an acceptable part of the shrub, it might have lost the odors needed to get males coming to it. Instead the bees settled on stems near site B, stems that had probably been used by some sleeping males the night before and so still had high enough concentrations of bee perfume to make them attractive to incoming males.

Time for another round of research. After the bees left on the morning of the 26th, I did another stem-switching experiment, starting with a new stalk where twenty-two males had bedded down the preceding evening. I moved this newly popular stem about a foot away, keeping it in the center of the shrub more or less, and replacing it with a stalk that I knew had not been slept upon recently, if ever.

The next night the popular stem attracted fourteen sleepers, even in its new location, while its replacement attracted no males at all. Hallelujah! These results supported my evolving explanation for the bees' sleeping choices, namely that they do favor stems that have been recently slept upon, provided that they are in the central portion of the brittlebush, rather than on its outer edge.

I planned some more "bed-shifting" experiments to check on the validity of this conclusion. My next experiment, however, was a complete bust because a shifted popular stem somehow managed to break overnight, as I discovered when I went to count the sleepers the next morning. Then, we had a heavy rain, after which the numbers of males returning to the brittlebush began to decline precipitously, so that by the 29th, I could find only a couple of survivors, not enough for an experiment of any sort. As a result, my conclusions about the basis for perch preferences must remain tentative. I await a new opportunity to watch *Idiomelissodes* sleeping in my front yard. At least, I am now prepared to test the hypothesis that popular stems must combine central position in a shrub with recent use as a sleeping site by the bees.

But having done our best for the moment on the issue of the cues that cause males to sleep together on certain stalks, we can now consider what benefits males might derive from their sleeping clusters. This question has been examined by several researchers who have studied sleeping aggregations of various bees (and wasps), not including *Idiomelissodes*. Most workers have wondered if by forming diffuse or not-so-diffuse associations, individuals in sleeping clubs derive some sort of protection against predators. There are, however, alternative hypotheses, including the possibility that males gain some thermoregulatory advantages by clustering. For example, if individuals form a ball, then bees in the center of the group might by protected from climatic extremes affecting those on the outside, although these bees might still be better off than those that were totally isolated. Or groups might form to find mates conveniently.

Neither the thermoregulatory nor the mate-locating hypothesis seems applicable to *Idiomelissodes*. First, although the bees often sleep touching one another, they do not form dense balls, and so cannot keep warm by using others as blankets, not that they are in any danger of becoming cold on summer nights in Phoenix. Moreover, because the sleeping groups are apparently composed entirely of males, we can confidently discard the idea that they are getting together to mate. Therefore, we will follow the herd and see if predators might be responsible for male sleeping clusters.

A complication here is that clumped males might experience several different kinds of the anti-predator benefits. First, according to the many-eyes hypothesis, if an approaching predator frightened one alert male in a group, its flight might warn others of impending danger. As a result, members of the group could escape more quickly than an isolated individual dependent solely on its own powers of predator detection.

But perhaps males in sleeping clusters practice active group defense, with aggregated bees assaulting an enemy more effectively en masse as opposed to going one on one with a predator.

A third possibility is that clustered individuals benefit simply because they dilute the risk of being an unlucky victim when they sleep together. Imagine that a bee-eating predator will kill one bee per evening, no matter what evasive response the bees take. In this case, it is better from a bee's perspective to be part of a cluster of 100 individuals as opposed to 20. Each bee in the mob of 100 has a 1 percent chance of dying per evening as opposed to the 5 percent chance of extinction for a bee in a lottery with just 20 participants.

Unfortunately, no one seems to have tested these three explanations for communal sleeping by male bees. On one July evening, I was lucky enough, however, to see one predator in action, an assassin bug, *Apiomerus flaviventris*, which buzzed into the brittlebush just as the bees began to assemble in the evening. The brilliant red, black, and yellow bug crash-landed on a dried brittlebush leaf and then stood with its front legs raised and spread, as if it were surrendering to someone. The front legs did not taper in the manner of a typical insect leg, but were thick and stumpy, giving the assassin bug the look of a double amputee. Had I chosen to inspect these limbs under a hand lens, I would have found them covered with short hairs that terminated in globules of glue, secreted by glands in the leg. The stickum helps the predatory bug hold on to prey, which it encloses in a tight embrace with those odd front legs.

As the waiting *Apiomerus* stood on the leaf, incoming bees occasionally approached it closely, to my surprise. They zigzagged erratically overhead or briefly hovered above the bug's back. The predator in turn shifted its stance on the leaf and twice lunged at a flying bee, trying to trap it between those powerful and sticky forelegs—but it missed both times.

After five minutes of foraging futility, the bug whirred off its perch. This time it landed at the base of a stem where about ten bees had lined themselves up for the night. The *Apiomerus* climbed the stem a bit clumsily, holding its forelegs in the pounce position. The bees ignored the rapidly approaching bug completely. It grabbed at the low man on the totem pole of sleepers. I held my breath. The male bee struggled mightily, and slipped the grasp of the killer bug. After the escape of potential victim number one, the *Apiomerus* tried again with the next bee in line, which had paid no attention to the life-and-death drama behind it. This bee, however, also managed to break free and so it went, one bee after another scrambling off, until the predator came to bee number eight. Finally, the assassin bug lived up to its name, holding firmly on to this bee with its forelegs and piercing it with a long, curved, stiletto proboscis. The "saliva" of assassin bugs contains a toxin that quickly immobilizes its prey, and the pierced male bee promptly ceased struggling, permitting the bug to leisurely suck up liquefied bee. Draining the prey required at least an hour, but much later that evening when I checked the bug by flashlight, I found it in a new spot with a new prey on its beak—liquid meal number two.

The next morning the husks of the two bees lay in depressions in brittlebush leaves where the predator had dropped them. The *Apiomerus* remained in view but had busied itself in a new and different task. It dipped its abbreviated forelegs into a bubble of fresh brittlebush resin welling out of a broken stem. The bug wiped freshly collected material from the tips of its forelegs to the upper part of each midleg. How odd. Was the bug gathering sticky resin to augment its own supply of stickum? If so, why wasn't it smearing the stuff on its prey-catching forelimbs?

Later Thomas Eisner, the same person who has analyzed the chemical defenses of earwigs, told me I had seen an *Apiomerus* gathering compounds that would help protect her eggs. The female transfers resins from plant to foreleg and then from foreleg to midleg and from midleg to the underside of its abdomen. After laying a cluster of eggs on a leaf, it then rubs its belly

Native
Stingers

.

91

on them, covering the eggs with a chemical blanket that repels insect egg consumers, particularly ants. Thus, resin collecting has nothing to do with prey capture and everything to do with egg protection.

To produce her eggs, female *Apiomerus* kill bees whose reaction to her tells us something about the possible reason for sleeping clusters. At no time did the males I observed mount a concerted attack on the predator, nor did the struggles of one escaping bee appear to alert its neighbor that an assassin bug was on the prowl in the immediate neighborhood. Indeed, once the bees settle down to sleep, a human can poke, prod, and paint them without eliciting much reaction, even from the molested individuals. The absence of a group response of any sort suggests that the many-eyes and group-defense hypotheses do not apply, at least when the predator is an *Apiomerus* assassin bug. But with just one observation, I cannot state with conviction that the dilution-of-risk hypothesis carries the day (or evening) in the case of sleeping male *Idiomelissodes*. I have to hope that the bees will return to sleep in our front yard during the summers ahead so that I can resume work on the mysteries that they pose.

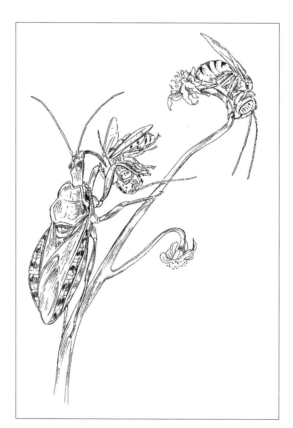

Apiomerus
assassin bug with prey

Earlier in the year, when the zucchinis stood upright, unscorched, and hyperproductive, males of the hairy squash bee *P. pruinosa* also faced the problem of how to survive the night in the outdoors. They did not have access to nest burrows, home only to females. Unlike *Idiomelissodes*, male squash bees opt for a solitary solution to the problem of finding a safe sleeping site. Males squeeze into a withered squash flower whose petals have collapsed and closed in on themselves, forming a small enclosed chamber where a squash bee can rest free from assassin bugs and other small terrors of the insect world.

A mildly solitary person myself, I'd prefer the male squash bee's solution to his sleeping problem if I were a male bee without a safe burrow or a stinger for defense. But male *Idiomelissodes* demonstrate that a social life is not a strictly female proposition in bees, wasps, and ants. Admittedly, the art of communal sleeping is a modest social skill compared to the intrigues that occur in the female social clubs of Hymenoptera. Which raises the question—why are male bees and wasps so uninvolved in their sexual partners' social lives? Many factors probably contribute, but, as Christopher Starr has written, perhaps "the sting's the thing." Starr speaks from personal experience, having written, "It has been my pleasure to provoke colonies of fifteen *Polistes* species in various parts of the world."

Stings require stingers, and in the bees, ants, and wasps, stingers are highly modified ovipositors (egg-laying tubes) that have evolved into syringes for injecting venom into creatures that have caused offense. All of which means that stingers and the ability to sting are a females-only phenomenon in the Hymenoptera. It is this ability that renders them a force to be reckoned with, particularly when females assemble in groups, as in paper wasp colonies.

Interestingly enough, among the termites both male and female offspring contribute equally to the sterile worker brigade, unlike the social bees, ants, and wasps, which have only female workers and subordinates. Unlike the hymenopteran ovipositor, the termite ovipositor is used strictly for its traditional purpose, laying eggs; nest defense has come to center on the jaws, and associated mandibular glands, which are structures that both males and females possess. The jaws of the soldier class of worker termites come in a great variety of shapes and sizes. There are slicing, crushing, or stabbing jaws; there are asymmetric jaws that fly violently apart, in a finger-snapping sort of motion; there are jaws that have fused into tubes through which the soldier can spray out strands of glue that entangle enemies. Because both male and female termites have the wherewithal for nest defense, both sexes have something useful to contribute to colony life.

In contrast to the termites, male paper wasps never make themselves useful to their nestmates. An attack by a mob of males would have no sting to it at all, which may make males largely irrelevant to the nest-defending societies of their sisters and mates. Freed from this responsibility and evidently unburdened by stinger envy, male Hymenoptera attend to their own agenda, which revolves around copulation rather than indiscriminately educating predators and innocent bystanders. As an innocent bystander, I can live with that.

CAMOUFLAGE EXPERTS

Everything that deceives may be said to enchant.

—PLATO
The Republic

THE COLOR OF my front yard differs from that of all other nearby lawns on Loyola Drive. The reddish soil and gravel, Pioneer Rock and Gravel's best, distinguish my plot from the somber gray pebbled desert yard across the street and from the varying shades of green grassy lawns up and down the street. Therefore, I marvel at the reddish desert grasshopper in my front yard, so beautifully color coordinated with my gravel that I only saw the hopper when it leapt away as I was about to step on it. How came this grasshopper to the only island of suitable habitat on Loyola Drive, so complementary to its color pattern, such an anomaly in a sea of Bermuda grass? The grasshopper is one that is common enough in the real desert miles away, but even there you won't find it sitting around on gray or brown gravel patches. It likes reddish gravel, period. And it is darn good at finding what it likes, so much so that it tracked down our place while traveling through Tempe.

Grasshopper watching offers modest entomological entertainment when walking in Arizona and elsewhere. Over the years I have kicked up a good many different species in the desert close to home, mostly those that jump hastily out of the way of a descending hiking boot rather than become flattened fauna. One reason why desert grasshoppers are so tough to spot lies in their skillful selection of just the right background on which to sit tight. Thus, the Arizonan species that routinely rests on fine white and gray pebbles in dry desert washes looks a lot like a dribble of little white and gray pebbles with a few pink ones thrown in for effect. Another species, a percher on rocks colored with a pointillist mix of black, gray, and tan dots, features a chaotic camouflage of small blobs of black, gray, and tan. In their turn, brown leaf-mimicking grasshoppers sit quietly among fallen leaves of the same color, whereas those that resemble red leaves naturally occur on the ground in the company of red, not brown, leaves.

It is easy to imagine what benefit a grasshopper might derive by sitting on a background that closely matches its own coat color(s). Many predators, especially keen-eyed birds, find certain grasshoppers most delectable. Edible grasshoppers almost certainly improve their chances of avoiding detection by their enemies if they prefer to perch on the appropriate background.

Background choice has not been studied experimentally in many grasshoppers, but work with one species is revealing. The species in question comes in a variety of color forms, a common phenomenon in this group

Cryptic gravel-
mimicking
grasshopper

The following photographs document the changes that occurred after I completed my herbicidal campaign and yard-scalping mission on a Kubota tractor during the summer of 1988. Here is the yard in the fall of 1988 after removing the Bermuda grass and depositing tons of Mission red gravel on the ground. A number of shrubs, including a small paloverde, have been installed and recently watered.

After three years, the paloverde has not grown a great deal, but spring wildflowers give the yard a less naked look during the months of March to May.

Two desert annuals close-up. I collected seeds of desert poppies from plants in the wild and mixed them with store-bought seeds of the Mojave desert bluebells. Cast among the gravel in the fall, they germinated after winter rains and bloomed in the early spring.

Penstemons also flower in the spring. These grew from seed produced by a handful of transplants purchased at nurseries in the Phoenix area. Some plants will last only a single year, but others are biannual. All are happy to refill the seedbank for the next season.

By 1994, some perennial shrubs had reached full size. In the middleground, the pink flowers of globe mallow; in the background, the bright yellow flowers of brittlebush.

Globe mallow flowers (at close range) attract a small army of native bees and wasps. Mallows are among the easiest plants to grow and most drought tolerant too.

Months after brittlebush has ceased flowering, one native bee uses its leaves and stems as sleeping platforms during the summer evenings. Thus, even nonflowering plants in a desert yard offer resources that attract native insects, whose presence can provide hours of entertainment for entomologically oriented observers.

The wonderfully aesthetic, if rather bland, fruits of Yellow Pear tomatoes, which the plants produce in superabundance.

I reserved a small but important part of the front yard for vegetables. The garden is bordered by desert senna, with the edibles shown (front to back) including summer squash, white-stalked Swiss chard, and Yellow Pear tomatoes.

By 1995, the front yard mimicked Sonoran Desert chaparral more convincingly thanks to the growth of a creosote bush (far right), paloverde (center rear), and butterfly bush (foreground), among others.

The now full-grown paloverde's springtime flowerfest feeds still more bees and butterflies, providing an always-growing list of subjects for front-yard entomology.

I have never regretted exchanging Bermuda grass for the wonderful diversity of
a Sonoran Desert landscape.

of insects. One color type is reddish, another green, and another gray. Grasshopperologists noted that the reddish form usually perched on the red granite rocks of its Coloradan home, whereas the greenish hoppers picked green vegetation and the gray ones favored gray shale as their resting site.

Interestingly, the insects choose perches by comparing the color of the cuticle around their eyes with the color of potential resting spots. Grasshoppers of the reddish persuasion that were painted green about their eyes by body-painting entomologists were much more likely to pick a green perch than were those individuals that had red about their eyes. Conversely, green hoppers that received the red eye treatment tended to go for red perches rather than the green ones they had preferred in the past.

The functional significance of color matching by these grasshoppers became clear as a result of some additional experiments in which individuals were either tied to or anesthetized and laid out on rocks of different colors. When gray hoppers were forced to take up residence on red rocks, lizards spotted and ate 82 percent of the unwilling participants in this experiment; in contrast, gray hoppers tied to or placed on gray rocks survived much better, with only 40 percent found and eaten in the same period.

Grasshoppers usually can move about, picking and choosing where to spend their quiet moments. There are, however, plenty of edible insects of relatively low mobility whose conspicuousness depends less on their own choices than on decisions made by their mothers. I speak especially of the caterpillars of certain moths and butterflies, whose adult females decide where to lay their eggs. The right choice will provide their larvae with a plant whose leaves they can consume and in which they can hide because of a correspondence between the color pattern of the caterpillar and its host plant. I encountered a fine example of this phenomenon in my backyard on a winter's afternoon while installing my bicycle in its resting place on the back patio. In the course of fiddling with the bike lock, I happened to look over at some nearby leaves of the Australian bottle tree that shades the bike rack. The leaves of Australian bottle trees hang in clusters suspended from long, supple stems that droop down from corky limbs and branches. The stem closest to me had a small twig projecting from it. I would not have given the twig a second look if it had been farther away or if it had been angling downward. But this twig was going against the grain, pointing upward, while attached to a bare stem, when every other twig, stem, and leaf petiole hung limply downward.

I went inside and retrieved my wife. Together we went back to the tree where I pointed to the leaf cluster and asked Sue whether she could find something interesting in the tree before her. She peered perfunctorily at the leaves and stems and gave up, used to this sort of task and having better things to do with her time. "Take a closer look at the twig," I urged. She complied a bit wearily but then exclaimed, "My gosh. It's a caterpillar." Which it was. To be precise, it was a caterpillar of the sort commonly called inchworms, which are produced by the many species of geometrid moths in the world.

Some inchworms possess an uncanny resemblance to the twigs of the plants whose leaves they eat. My bottle tree inchworm had skin finely lined with brown and green, giving it a smooth tan look that more or less matched the color of the bottle tree stem. In addition, the caterpillar's head was abruptly flattened in front, not rounded, the better to look like the blunt

end of a twig. Near the tip of its abdomen was a small projection, like an irregularity in the bark of a twig. The final touch was a false leaf node, a crescent-shaped projection about one-third the way down the caterpillar's back that looked like the bump left behind on a twig when a leaf falls from it.

Despite the caterpillar's color and body shape, its twig mimicry would have been useless had it not also behaved like a twig, holding its body ramrod straight, angling away from the stem at about 30 degrees. It grasped the twig perch with two pairs of false feet on the end of its abdomen; a fine filament of silk looped from its flat-faced head around the stem. When Sue and I peered at the inchworm closely, it did not flinch but remained absolutely frozen in its twig-like mode.

I went out at eight that evening, armed with a flashlight to check the inchworm, only to find it away from that afternoon's resting place and no longer "afflicted" by apparent rigor mortis. Instead it was on the move, humped up into an inverted U in a leaf cluster well below its old perch. At 10 P.M., it had inched a short way and was now stretched out gripping a leaf petiole with its hindfeet while its head lay close to a leaf edge. The leaf edge had been nibbled, creating an irregular depression in the tissue. At seven the next morning, the caterpillar pressed close to the stem, several inches above the nibbled leaf. By the time I returned to park my bike that afternoon, it was in the twig position on the open stem about where I had spotted it the day before, a good foot from its nocturnal feeding place. That evening after dark it again migrated from its daytime resting spot to feed on a leaf well below it, turning around before dawn to backtrack to the spot where it would once more assume the alias of a twig.

I had several questions for the twig-mimicking inchworm. First, why had it made the "mistake" of perching so that its body formed an upward-pointing twig on a plant whose real twigs point down. Even a shortsighted entomologist could spot it, never mind a visually unchallenged bird predator. One possible answer is that the caterpillar was *not* an Australian geometrid whose ancestors had traveled with Australian bottle trees during shipment to nurseries in the United States for purchase by our landscape gardeners and developers. Presumably any moth that had evolved in close association with Australian bottle trees would produce caterpillars whose color and behavior were perfectly matched to the host plant.

Typically, the color match between twig caterpillar and host plant is indeed darn near perfect. For example, one species that I have found on the tough little oaks in the Chiricahua Mountains of southeastern Arizona has a slightly mottled grayish skin color and assorted bumps and protrusions that make it a dead ringer for the gnarly twiglets of the oak, particularly since the caterpillar aims itself on stems in the same direction that real oak twigs are pointing (toward the tip of the stem).

In contrast, our backyard twig caterpillar not only faced in the wrong direction when perching but also had a color pattern that only approximated the color of bottle tree stems. So I guess that this caterpillar belonged to a species of geometrid native to Arizona whose mother happened to lay an egg or two on a novel food source, an Australian bottle tree. The larva was clearly capable of consuming and digesting bottle tree leaves and its twig mimicry provided it with respectable, although imperfect, camouflage. Of course, I cannot completely eliminate the possibility that the inchworm actually was just an odd specimen of a fair dinkum Aussie species that happened to perch against the grain.

Leaving this issue, we can turn to another small oddity—why doesn't the caterpillar feed all day long? Surely the insect would gain weight more quickly if it fed around the clock. It is not hard, however, to imagine why inchworms may have to forgo activity during the day because it is then that insect-eating birds search through the foliage for their victims. A bird's visual system is remarkably sensitive to movement, rather like our own. Immobile twig mimicry may help some geometrid larvae avoid detection and the hard snap of a sparrow's or warbler's beak.

The great ethologist Niko Tinbergen tested this possibility in experiments with hand-reared European jays. He and his colleague de Ruiter released their tame jays one by one into an aviary on whose floor they had scattered twigs from birch trees and the twig caterpillars that live in and on birches. The first test jay behaved like most of the later participants. The bird hopped about the aviary looking around in the alert and curious manner of jays until it happened to step upon a twig caterpillar, causing the caterpillar to twitch. The larva's understandable reaction proved nonetheless a mistake. The jay promptly grabbed it with its beak, pounded it on the ground, and swallowed it with gusto, despite having had no previous experience with twig caterpillars.

Since the jay had not been fed for some time before the experiment, just one caterpillar didn't do the trick. It set out to hunt for more. Over it hopped to a real birch twig, which it picked up, bit firmly, and dropped.

After several such unrewarding encounters, the jay subsequently ignored all other twigs and caterpillars alike. Further experiments with jays and other insect-eating birds confirmed the initial finding that at least some birds have a hard time separating twig-mimicking caterpillars from the real McCoy, but only if the twigs were from the food plant on which the twig mimics lived (and which they closely resembled).

Thus, being a nonmoving twig mimic has a survival payoff that probably compensates the caterpillar for its loss of food during the daylight hours. Still we are left with a final puzzle: Why should a twig mimic travel a foot or more to reach a daytime hiding place when there are plenty of good resting sites a few inches away from its next meal? Bernd Heinrich was inspired to study this problem, having discovered a whole catalogue of caterpillars commuting between a nighttime feeding area and a daytime retreat. Heinrich suspected that the larvae of these species were trying to distance themselves from leaves whose shape they had altered with their bites. He guessed that if leaf-eating caterpillars were to stay by leaves they had damaged and if caterpillar-eating birds were smart, the birds would need only to scan the foliage for distinctive scalloped irregularities to find the next meal.

To test this hypothesis, Heinrich and a coworker, Scott Collins, offered captive chickadees a chance to forage in living trees contained within a large aviary. The two men altered the leaves of some trees using a paper punch to mimic caterpillar feeding damage. To the damaged leaves, the experimenters attached mealworm bits, a food dear to hungry chickadees. Birds permitted to enter the aviary proved to be quick learners. In short order, they made the association between leaf damage and snacks, and over a series of trials, they became increasingly adept at rapidly depleting the supply of mealworm McNuggets. Case closed.

But, wait. There are many different ways to go about things and not all edible caterpillars hide from their predators this way. Some highly conspicuous species avoid looking like something to eat and so avoid their consumers. I found a wonderful case of flamboyant camouflage on the citrus trees outside a bedroom window.

When I turned the front yard into a sliver of desert, I left intact the trees that border the other three sides of our house. These untouched medium-sized trees include an African sumac, a Chinese (or Nepalese) mulberry, an Australian bottle tree, an Australian eucalyptus, and a row of domesticated citrus whose wild ancestors originated in Southeast Asia. Later, variants were imported to Spain by the Moors, and from there they were introduced into the West Indies and South America, where still other forms were developed and subsequently transported to what is now the United States. Although all our yard's trees are indisputably exotic in their origins, they are a considerable cut above Bermuda grass and red brome. Even the mulberry, which generates a formidable amount of pollen in the spring, has some redeeming social value thanks to the summer shade it offers as compensation for spring sneezes. I haven't summoned the necessary resolve to woodchip the allergenic mulberry and its foreign companions and replace them with native ironwoods or mesquites. But I keep hoping that the mulberry will die soon of its own accord.

Even if the mulberry does succumb to old age or disease, I will continue to spare the citrus, one lemon, one tangelo, and one grapefruit, which stand up bravely to the sun on the west-facing side of the house and provide us with great quantities of fruit from January to May each year. They do require deep watering every two weeks or so, but since they repay us with edibles, I am inclined to treat them as extensions of the vegetable garden, entitled to protection from the Native Plants Police and to their fair share of Salt River water and TipTop fertilizer.

Because of their honored status as agricultural plants, I keep an eye on them, checking for wilting leaves or fruit drop or thrips. In the course of one such check, I noticed a small object draped half over a citrus leaf, half over the adjacent petiole. "Oh," I said to myself absently, "a bird dropping." Or something to that effect. And in the same thought, I added, "Hold on," when I suddenly realized that I was looking at a butterfly larva, not a bird poop. And what a caterpillar! Of just the right size (about three-quarters of an inch long) to be a cylinder of bird dung, the larva's color pattern beautifully advanced its humble mimicry. Its outer surface combined asymmetric patches of green, white, and deep brown of the sort anyone who is so inclined can admire in a typical bird dropping. Moreover, the whole thing fairly glistened as if still moist, fresh squeezed out of a cloaca and plopped onto the leaf by an English sparrow or mockingbird that had paused briefly in the shrubbery and then had gone on its way unburdened.

The complexity of the caterpillar's dung mimicry, the interplay of size, shape, color pattern, the illusion of moistness, and the larva's droopy resting posture, quite took my breath away. And I am not the only one. One

Dung-mimicking
caterpillar of the
giant swallowtail
with adult butterfly

professorial odd job is to answer questions from the public when they call the university in search of information. Since I am supposed to be an "insect specialist," calls about bugs are often relayed to me. Just the other day I was connected to a person who asked whether I could identify a caterpillar for her based on an over-the-telephone description. I replied that there might be one chance in fifty but to go ahead. She told me that she had been out pruning her orange tree when she noticed a large bird dropping on a leaf—and then it moved! Only then did she realize that her tree had been harboring a caterpillar whose deception she found remarkable.

I was delighted that my initial pessimism had been unwarranted and I told her that she had discovered the larva of the giant swallowtail butterfly, *Papilio cresphontes*. The news seemed to please her, particularly when I assured her that the creature would not seriously damage her citrus tree and that she could, if she wished, rear a gorgeous butterfly from the beautifully ugly larva by providing it fresh leaves and a stem on which to pupate. It is always gratifying to do something useful when one is an academic.

But returning to academic issues for a moment, why should this caterpillar have gone to the trouble of perfectly resembling something as pedestrian as a bird turd? One could be forgiven for thinking that the existence of this insect is evidence that the Creator has (had?) a not terribly sophisticated sense of humor. Let me advance, however, another less perverse explanation for the bird-dropping caterpillar—its mimicry helps protect it against caterpillar-eating birds. One imagines that insectivorous birds scanning the leaves of a tree for consumable insects might well pass quickly over bird droppings—and things that look like them—since avian excreta are not in the diet of insect-hunting birds.

This proposition could be tested by presenting captive insect-eating birds with an opportunity to forage in a cage that contained some bird-dropping caterpillars as well as some caterpillars of the same size but decorated with a different color pattern. We would expect the dung mimics to be less frequently picked up and eaten than the other prey.

This experiment has never been done, in part perhaps because it would be difficult to find an appropriate alternative caterpillar, one as conspicuous as the dung-mimicking larva but not protected against predators in other ways, notably by poisons in its tissues or by stinging spines or hairs.

However, we can test the dung-mimicry hypothesis in another way. The putative dung-mimicking caterpillars of the giant swallowtail change size dramatically as they proceed from the tiny hatchling just out of the egg to the nearly thumb-sized final stage. I do not know what your experience has been but I have not encountered bird droppings the size and shape of a man's thumb sprawled out on the leaves and stems of trees and shrubs here or elsewhere. If my experience is typical, a thumb-sized caterpillar that looked like a bird dropping would fool no one, least of all intelligent birds that like caterpillars for lunch. Therefore we would not expect the larger larval stages of *P. cresphontes* to resemble bird dung—and they do not.

Instead, as the caterpillar grows, it molts, exchanging one color pattern for another along with a more expansive cuticle. These larger stages have a color pattern that features two large patches of dark chocolate separated by a saddle of white. The abrupt changes in color help disrupt the outline of the caterpillar, making it appear like a thick dark twig with a patch of pale lichens growing on it. Moreover, the caterpillar in this metamorphosis possesses an osmeterium, an inflatable, strong-smelling gland that pops out of the dorsum of the caterpillar near its head, should you molest it. Just what category of natural molesters the osmeterium and its chemical constituents are designed to deter is a matter of debate. However, in some other swallowtails, the larva's osmeterium does secrete quantities of isobutyric acid and 2-methyl butyric acid, which the caterpillar smears on attacking ants, much to their discomfort. Whether these chemicals also deter any vertebrate enemies is apparently unknown.

Large caterpillar of the giant swallowtail with osmeterium everted

In addition, if the smaller dung-mimicking stage of the giant swallowtail really does fool some creatures into thinking it is bird doodoo, then other cases of apparent dung mimicry by insects should involve species that are neither too big nor too small to be convincing bird deposits. Moreover, the presumptive dung mimic should spend long periods immobile on the upper surface of leaves, just as the dung of arboreal birds does.

In the course of my wanderings, I have encountered many creatures of this sort, including a small chocolate-and-white moth that perches during the day on the upper surfaces of leaves with its wings folded in such a way as to create a thin turd-like cylinder. The result is an object of just the right size and shape to masquerade as a bird dropping. Likewise, the small larvae of the European alder moth (which I have not seen in the wild) beautifully resemble black-and-white bird droppings; but then as the caterpillars grow larger and become less able to mimic dung persuasively, they switch to a striking black- and yellow-banded color pattern, which may warn predators of poisons within.

Similar, but even more diverse, switches occur during the growth of caterpillars of *Oxytenis naemia*, a Panamanian moth. After hatching and during the early stages, the little larvae look like small black bird droppings. As the caterpillar reaches an intermediate size, it takes on a different color pattern mimicking a pale, shiny dropping direct from a bird of more substantial size. Finally, after undergoing yet another molt and achieving a length of about two inches, the now really big caterpillar abandons all imitation of avian feces and instead resembles a rolled-up dead leaf. Should you touch the caterpillar in its leaf-mimicry phase, it expands its head, revealing two false eyes, and turns sharply toward its molester, as if it were a snake striking back. Clearly, this caterpillar covers all the bases.

Potential dung mimicry is not the sole province of butterfly and moth larvae inasmuch as a number of tropical long-horned wood-boring beetles also employ this fakery. They achieve a resemblance to dung by adopting weird postures that give them a most unbeetlelike asymmetry, a shape that transforms them into a typical haphazard pile of bird feces. For example, adults of an odd mottled brown beetle lie on their side with two lumpy hindlegs held together and pressed on or near a leaf, making the insects look for all the world like arcs of seedy excreta cast out by passing birds. In another species, the beetle sticks out one whitish foreleg, but not the other, thereby disrupting the visual symmetry of its body that might give its true identity away. Both beetles are highly reluctant to move, even when poked and prodded, as behooves an object that is supposed to be bird dung, which rarely wiggles when approached by a predator.

One need not visit the tropics to see beetles masquerading as a mound of bird dung. On my parents' farm, I have found an immature leaf beetle that retains its own feces and the cuticle of cast skins each time it molts, forming a sheet of brown, granular material that it holds over its back. Viewed from above or from the side, one sees only a small pile of what looks remarkably like a modest bird crap (at least to a human observer). The deception breaks down only when the beetle moves, which it rarely does during the daytime while resting in full view on the upper side of a leaf.

There is an alternative, but not mutually exclusive, explanation for the leaf beetle's eagerness to retain its feces, which is that the insect can employ the stuff as a shield against biting ants. Thomas Eisner and his associates have shown that the beetle larva reacts to an attacking ant by interposing the firm mass of feces between its soft body and the steely jaws of the ant, often with a result gratifying to the beetle—namely, the departure of the ant. Whether the larval chrysomelid also uses the fecal shield to "hide" from bird predators is unknown. As indicated, many potential cases of dung mimicry await attention from entomologists, who may feel that the subject is not sufficiently elevated to warrant a large investment of time and energy.

Although I have not rushed in to fill this entomological void, perhaps I will someday, comforted by the knowledge that Charles Darwin spent several years engrossed in the study of earthworms and their castings, an equally unprepossessing subject, or so it would appear on first glance. But just as Darwin's obsession with earthworms led to advances in understanding landscape changes and soil formation (he showed that the battalion of British earthworms living in a typical acre transported about ten tons of soil to the surface each year), I can envision how a thorough analysis of dung mimicry in insects could contribute in a modest way to a big issue in biology, namely how color pattern and defensive behavior evolve.

Such an analysis would deal with many little-understood issues, including why the young larvae of the giant swallowtail and some others employ their particular dung-mimicking stratagem whereas the larvae of many moths, members of the same order (the Lepidoptera) as the swallowtail, mimic twigs to an extraordinarily accurate degree, while still other lepidopteran larvae employ stinging hairs or communal warning displays or various other means of deterring their predators. In other words, if dung mimicry is effective for certain sizes of giant swallowtail larvae, why isn't the tactic used by many more butterfly and moth larvae of the appropriate size? Could it be that the tactic only works reasonably well if there are relatively few bird-dropping mimics in a region? Too many, and perhaps the local birds might catch on, even to the point of hunting specifically for anything that looked like bird poop. But if there is a limit, when is it reached? Getting a real handle on why a caterpillar of species X is a shiny dung mimic while species Y looks like a tan twig would be satisfying indeed. It might even compensate a researcher of dung mimicry for the inevitable jibes about his or her state of arrested development. So bring on the bird-dropping caterpillars. Let's get this thing started.

ALIENS

And so the greed continues, I imagine until mankind eventually becomes extinct and the next species takes over the Earth, the insects perhaps.

—EDWARD F. HORST
in a letter to the *Arizona Republic*

BIKING UP TO my on-campus office today I am bombarded by a stupendous number of insects. Is this what Mr. Horst meant when he contemplated the takeover of our planet by bugs? A blizzard of tiny white flakes drifts through the air, creating a virtual snowstorm, although it is early September and the day is headed toward 105 degrees.

I know what the white bits are, having made their acquaintance in my vegetable garden, where they line up shoulder to shoulder on the leafy undersides of squash, beans, broccoli, and almost everything else at this time of year. They are whiteflies, plant parasites extraordinaire, members of an order of insects, the Homoptera, known to entomologists as "true bugs," so small that even a nearsighted person has to peer intently at a specimen to make out its pale white wings and spindly little legs, which power feeble attempts to walk through the thicket of hairs on my forearm.

When I try to brush the whiteflies from my arm or shirt, I mangle and crush them by the dozens, so fragile and soft-bodied are they.

And yet these insects, despite their apparent insignificance and extreme vulnerability, have so impressed agricultural circles as to be crowned "superbug." Their respectful title comes to them in part by virtue of their ability to remove fluids from, and simultaneously transmit deadly viruses to, about 600 species of plants including many economically important ones, such as cotton, melons, squash, and a host of other vegetables.

Like other more familiar true bugs, such as aphids, the whitefly comes armed with long, thin, piercing mouthparts, which are inserted into plant tissue. The nutrient-containing phloem of the plant flows up the tube-like mouthparts and into the whitefly's digestive tract. Certain valuable substances in the fluid are captured and utilized by the bug, while the excess liquid continues on, soon to be dumped out unused. When one thinks about it, the ability of whiteflies, aphids, and their relatives to flourish on a diet of very dilute sugar water is astonishing. Plant saps are nutritionally deficient in most respects, and particularly short of protein. Yet whatever shortcomings are inherent in their diet don't deter sap-feeding true bugs, as demonstrated by the clouds of whiteflies drifting through the Arizonan atmosphere today.

As it turns out, most true bugs have solved the potential problem of nutritional shortfalls by forming a mutually beneficial arrangement with a variety of single-celled organisms. These "endosymbionts," many clearly

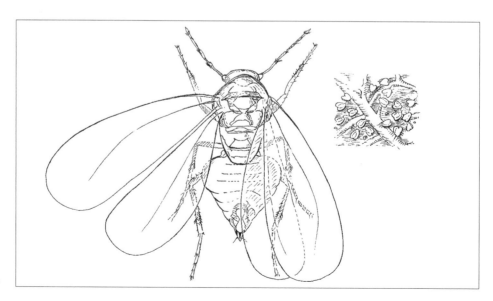

Superbug whitefly

bacterial in origin, have taken up permanent residence in special cells inside their aphid or whitefly hosts. There they lead a pampered existence in a hospitable cellular environment in return for producing and releasing metabolic by-products of utility to the bug that shelters them.

Among the key substances generated by endosymbiontic bacteria are certain amino acids not available in plant saps. They may also throw in some scarce vitamins as part of the bargain. The importance of these house bacteria to their host bugs was revealed experimentally. When aphids were given a dose of antibiotics in their diet, the antibiotics killed the resident endosymbionts, resulting in reduced growth, sterility, and the premature demise of aphids so treated.

Mother aphids of many species have elaborate mechanisms for transfer-ring symbionts to their eggs or to their offspring at the moment of birth, further evidence, if any were needed, of how hosts value their endosym-bionts. Such interdependence suggests a long and happy association, a conclusion checked through an analysis of the genes of the microorgan-isms encapsulated in bugs with those of their closest free-living relatives. According to these studies, some bacteria colonized certain homopteran guts and body cavities at least 80 million years ago, and probably millions of years earlier still.

The good work endosymbionts perform for whiteflies and the like does not release the insect hosts from having to process a huge amount of plant sap to secure what they and their bacteria-in-residence require. As a result, true bugs often ingest vastly more fluid than they can contain. Liquid waste moves smartly front to back and out into the wide world. In the case of whiteflies, excreta are neatly deposited into a pocket on the *upperside* of the abdomen. Here is an anatomical oddity. Why not void the unwanted fluid directly onto the ground from an anus on the abdomen tip, a system that works well for the vast majority of insects?

The answer to this question involves understanding that whitefly excreta, known euphemistically as honeydew, contain a lot of excess plant sugar that the bug did not sequester and use. As a result, whitefly honeydew is slightly sticky, and it becomes more so as water evaporates from it. The bug that dumped undigested honeydew onto the leaf where it and its sessile offspring live would soon be standing in a puddle of glue. This dilemma is made more severe as a result of the sheer volume of plant juices processed by the bug, which imbibes constantly while attached to its host. So the trick for the whitefly with its beak stuck into a leaf is how to keep from drowning in its own waste products, a problem with a certain resonance for modern humans.

It is to this end that the whitefly has evolved its unusual dorsal reservoir and an associated structure called the lingula. After enough honeydew passes out of the anus and into the reservoir, the whitefly cocks its lingula, depressing the rod-like structure into the fluid. When this device springs back to its uncocked position, it flicks the honeydew far from the bug, reducing the risk that the insect will become glued to the plant as a side effect of its fondness for phloem.

However, when huge numbers of whiteflies aggregate on a victimized plant, one bug may catapult a honeydew droplet onto another bug beneath it. Indeed, the collective result of a small army of whiteflies is to coat the entire plant and surrounding soil with the sticky residue of their excretions. A film of congealed honeydew probably clogs leaf pores and so interferes with the gas exchanges vital to photosynthesis. Moreover, the substance provides a congenial medium for sooty mold fungi, which compound the damage done to the afflicted plant by the whiteflies themselves as they siphon off valuable sugars, nutrients, and water from their host. These effects become particularly manifest when a plant plays host to a gazillion whiteflies, the sum of whose attacks becomes highly damaging despite the trivial contribution of any one parasite to the total account.

Mass attack by whiteflies on the rampage is, however, a relatively recent phenomenon. Only since the 1980s have their populations exploded in California and the Southwest, with spectacular increases in crop damage as a consequence. Melon production has fallen by 50 percent in some areas. Losses in cotton production caused by insects quintupled (from 2 percent to 10 percent of cotton output) between 1988 and 1992 during the rapid buildup of whitefly numbers.

Before the 1980s, a bicyclist in Tempe would not have passed through a swarm of whiteflies on the wing in September. In fact, only in the past four or five years have they turned their attention to vegetables, in addition to cotton, in a big way. These days broccoli and cauliflower sets planted in the late summer attract a standing-room-only crowd of white-winged pests. David Byrne and Jacquelyn Blackmer have found literally hundreds of immature whiteflies on single leaves from unfortunate melon plants at the height of an infestation. They also once counted more than a thousand whitefly eggs on a single square inch of broccoli leaf growing in Arizona.

With the coming of superbug, I first tried to fight back, putting out yellow cardboard squares daubed with Vaseline. Whiteflies love the color yellow, even pursuing yellow tennis balls rolling slowly across a tennis court, and as expected, they landed in squadrons on my homemade traps. I also sprayed the leaves of plants with water or with insecticidal soap, knocking other whiteflies to the ground, killing them right and left. Despite the slaughter, I only freed up leaf space for replacement whiteflies. For every deceased bug, there seemed to be a hundred ready and willing to fill in, many probably coming from nearby cotton fields when the farmers stopped irrigating them in anticipation of the cotton harvest.

Wherever the whiteflies came from, they came by the billions, sailing through the air, carried by gentle breezes that deposited some on a suitably edible landing place in the garden. Although whiteflies give the impression of being marginal flying machines at best, they are surprisingly competent at staying aloft, once airborne. In laboratory flight chambers, some individuals (admittedly not many) manage to fly for more than two hours. In the real world, bugs that decide to move out generally do so in the early to mid morning, taking advantage of ascending convection currents that form as the sun warms the earth. Although most whiteflies travel only a mile or two at most, some bugs carried by these currents have been found 5,000 feet up, floating along, saving their energy while traveling to far-off destinations.

But why should whiteflies have become so conspicuous and troublesome in recent years? Previously they hardly bothered suburban gardeners and farmers alike. To figure out what happened, let's first identify the species responsible for all the agricultural misery. As it turns out, of the roughly 1,200 named species of whiteflies, only a handful wander about reducing crop yields and giving their group a bad name. Public enemy number one is a beastie called *Bemisia tabaci*, the sweet-potato whitefly, although as we have seen, the bug most definitely is not a strict sweet-potato junkie.

Early alarms about *Bemisia tabaci* include a report from Greece in 1889, where the bug was a pest of tobacco, and an account published in 1905 about cotton damage in India. Since that time, much attention has been devoted to this species by economic entomologists, so that, as insects go, *Bemisia tabaci* is relatively well known to its entomological "admirers."

When some researchers took a look at the bug that was rampaging through California and Arizona in the early 1990s, they felt that it was slightly different in appearance from the most familiar form of the species, which was labeled "strain A." They called the terrorist whitefly "strain B" and thought it might be a new foreigner, accidentally transported into the southwestern United States where it could wreak havoc in an area lacking its natural biological enemies.

A team of entomologists subsequently reported in the prestigious journal *Science* that "strain B" was indeed a new species, which they called the silverleaf whitefly (*Bemisia argentifolii*). They based their conclusion in part on genetic differences between strains "A" and "B." Moreover, although the two types would court one another, females of strain B refused to copulate with males of strain A. Most biologists define species as populations that are reproductively isolated and genetically distinct. Therefore, these specialists felt justified in giving the two kinds of whitefly their own specific names.

However, trying to figure out what are separate species as opposed to geographically distinct races or forms of the same species can be a devilishly difficult task, especially when working with populations of very small, very similar insects. Differences of opinion on these matters are routine. In fact, in 1957, some entomologists condensed what had previously been considered nineteen distinct species in the genus *Bemisia* into a single species, *Bemisia tabaci*, of nearly worldwide distribution. Against this backdrop, the decision to split off *Bemisia argentifolii* was bound to generate debate. Many whitefly experts believe that "strain B" is just one of about fifty different "strains" of *Bemisia tabaci*, each population differing somewhat in genetics, reproductive behavior, and other attributes, but not so much so as to warrant identifying any one strain as a separate species.

Whatever you call strain B, or the B-biotype, it is no run-of-the-mill bug. The appearance of strain B in the Southwest caused my vegetables and me considerable distress, while profoundly saddening the region's cotton farmers and agribusiness operators, many of whom continue to suffer major losses each year from whiteflies.

But despite all the attention that has been lavished on strain B following its emigration to Arizona and California, no one knows for sure where it came from. This knowledge might well be helpful in attempting to control the little monster. For example, if its native home could be identified, whiteflies in that area could then be examined closely to see if they were held in check by certain local parasites or predators. If so, these insect enemies might be introduced into the Southwest where they might possibly help reduce the annual population explosions that make superbug such a super agricultural problem.

This claim may not be purely wishful thinking because infestations of the exotic ash whitefly, a distant relative of superbug, have been largely controlled by a ladybird beetle, a natural enemy of the ash whitefly in its native homeland. The ash whitefly made its way to California in 1986 or 1987. The ladybird beetles came from Israel courtesy of an Israeli entomologist, Dan Gerling, who collected them from a tree in his backyard. In Israel the ash whitefly causes little emotional or economic damage, thanks to these predators, which have done excellent work since being released in California in 1990. Although the beetle is not as extreme a specialist as the scale insect eater, *Rodolia cardinalis*, it has a special fondness for whiteflies, with female ladybirds capable of downing 10,000 whitefly eggs during their adulthood.

However, importation of a foreign predator of superbug may not work, since superbug may not be a foreigner. Instead, the bug may be a Frankenstein-like creation of our own agricultural practices. Wherever and whenever there has been a whitefly problem, one finds the same pattern: intensive production of huge fields of a single crop, like cotton, coupled with liberal applications of water, fertilizer, and a battery of pesticides. Even though we are now definitely in the post–Rachel Carson era, cotton farmers are still great aficionados of poisons, which they employ in a heavy-handed attempt to subdue a catalogue of cotton-consuming insects, of which the boll weevil is only the most famous. The typical field is sprayed at least ten times per season. Some operators shell out for as many as thirty sprayings of a single cotton crop at the cost of hundreds of dollars per acre.

The drenching of cotton fields with the pesticide of the moment has naturally favored any insect pest that happened to have a biochemical answer to the toxin. All the standard cotton pests have evolved resistance to all the major pesticides, and so have whiteflies, even though initially they were not a target of the chemical barrage. But once a mutant whitefly or two appeared that could cope with a particular poison, those individuals had a reproductive field day in fields treated with that substance. The various tiny parasites that normally infect and kill most whiteflies had been removed by repeated pesticide application, enabling the resistant white-fly pioneers to reproduce freely, which they do with abandon, with the B-biotype multiplying about five times as fast as other strains. As a result, populations of whiteflies have grown without restraint to their present frightening levels.

Needless to say, the search goes on for new pesticides fatal to whiteflies. One promising substance has very recently been extracted from an Andean plant; the chemical, a naphthoquinone, is said to cause whiteflies to go belly-up with gratifying speed, while supposedly leaving helpful insects, like ladybird beetles, unaffected. However, even if true, such a substance will be years getting to market. And when it arrives, spraying will proba-bly select for any mutant whiteflies that happen to be resistant to naphtho-quinone's generally deadly effects.

In the meantime, why not look for help from other things besides pesti-cides? Perhaps by putting toxins on the shelf, farmers might enable the var-ious minute enemies of whiteflies to recover, leading to a sharp increase in the proportion of young whiteflies that are handicapped by parasites. Par-asitism rates do indeed go up promptly in whiteflies after the spraying stops in cotton fields. If there were some way to convince cotton farmers to stop relying on the pesticide crutch, perhaps the whitefly menace might crumble under the weight of its own natural enemies. If so, fall in Arizona would be much more productive for front-yard gardeners.

Until this unlikely event occurs, however, suburban agri-operators can only wait out the summer months hoping for the arrival of cool fall weather. Chilly evenings come slowly to central Arizona, but when they do slide in, the clouds of whiteflies gradually dissipate, bicyclists in Tempe can open their mouths without inhaling superbugs with every breath, my front-yard broccoli plants can finally regain control of their destiny, and that spells R-E-L-I-E-F, to parasitize the old Rolaids jingle.

THE INTRODUCED WHITEFLY is only one among hundreds of animals and plants that now call Arizona home but once lived somewhere else. In the United States, more than 2,000 exotic insect species and an equivalent num-ber of alien plants have become established. There are over 300 species of introduced plants in Arizona alone. As I noted earlier in the case of cottony scale insects, some accidentally introduced organisms have a way of getting out of hand and causing economic havoc. This reality has not stopped people from purposefully engaging in wholesale translations of species from one place to another.

For example, dozens of grasses have been brought to Arizona from around the world to feed rangeland cattle or to stabilize soil in badly eroded areas that have been damaged by overgrazing. The various uninvited grasses in my front yard are mostly species whose real home lies far from the Sonoran Desert of central Arizona. I don't know which of the two species of *Schismus* I have in the front yard; I may be living with *S. barbatus* (introduced as forage for cattle from an Old World site), but perhaps I am blessed with *S. arabicus* (introduced from western Asia). Without question, red brome (*Bromus rubens*) is on board, and this introduced European grass tries each spring to take over completely.

According to *Arizona Flora*, young shoots of red brome are "relished" by cattle, but after the grass has reached maturity and has a foxtail head composed of extremely tough, sharp-pointed seeds, cows lose enthusiasm for the plant. The ability of red brome's seeds to penetrate a cow's mouth, a dog's feet, or a hiker's socks is notorious. (Friends of mine who have a large and rambunctious dog recently paid their veterinarian several hundred dollars to remove brome foxtails that had screwed their way far down the ear canals of their pet.) Those seeds that avoid going through a cow's gut or a washing machine or a veterinarian's office lie in the desert soil for months, waiting for even the most marginal of winter rainfalls to set them off again for a spring appearance. They can germinate on a mere half inch of precipitation, whereas the native desert annuals require at least an inch of rain. Thus, red brome elbows its way past the native species, and is now one of the dominant plants of the real desert as well as suburban desert landscapes that have not been sprayed with a chemical designed to block seed germination.

In addition to its outright ruthlessness, red brome can change the desert in another way, namely through the promotion of summer wildfires when dried brome offers abundant fuel for careless or pyromaniacal humans. In the past, the Sonoran Desert lacked a dense grassy ground cover. Fires were rare. As a result, the perennial native flora, the saguaro and barrel cacti, the paloverdes and ironwoods, lack the capacity to come back after being burnt. Years after a killing wildfire, a once-scorched patch of desert will still lack the slow-growing cacti and small trees that once grew there. On the other hand, red brome is an annual that produces seeds inured to fire. When these seeds germinate the next spring, the grass will sprout in open soil, now enriched by the nitrogen and other elements derived from burned plants. Brome swarms over old burns. When this thick grassy coat dries during the droughty months of late spring and summer, it offers new opportunity for even hotter fires. If some jerk does set the grasses alight, the new fire will finish off whatever perennials managed to survive the preceding blaze, creating conditions that favor only the annuals, especially red brome. Out goes the complex, three-dimensional, highly diverse chaparral of the desert and in comes a unidimensional, uninteresting prairie dominated by an introduced weedy grass.

So it is that I delight in dispatching red brome. As noted, my goal has been to suppress the weed through weeding, rather than by herbicidal attack, which would injure the native plants whose presence I desire. Therefore, I spend hours each spring ridding the yard of this pestiferous grass by pulling it up, one plant at a time, good exercise of a sort, but a Sisyphean task if there ever was one.

MY FRONT YARD also supports an Old World insect that was purposefully introduced into the Americas for its beneficial economic effects without a worry otherwise. I speak of the honeybee, which apparently came to the New World from Europe via Spanish colonists long before anyone knew that introductions had their downside. The positive products of honeybees are well known. Honey itself currently is a relatively minor gift from the insect, given that the agricultural crops pollinated by them are worth on the order of $10 billion in the United States.

But the honeybees that zoom about the front yard, collecting pollen from the zucchinis, visiting penstemons for nectar, and pollinating the citrus trees in season, are not an unmitigated blessing, especially from the perspective of the native bees of Arizona. Researchers working in New York forests have found that a typical honeybee colony removes per annum about 50 pounds of pollen and many more pounds of nectar from a considerable array of flowering plants growing within three miles of the hive. To secure their annual haul, thousands of foragers in the typical colony rack up some 15 million miles of foraging trips. That is a formidable gleaning effort and one that seems certain to make life much more difficult for native bees searching for the same resources. The possible decline of the generally inconspicuous native bees has not caused concern among the public here. However, in other regions of the world where honeybees have also been introduced by European colonists determined to have their honey, the competitive effects of these pollen pigs have attracted some outspokenly negative comment.

In Australia, for example, some naturalists think that honeybees crowd out the native pollinators, including certain of our fellow vertebrates. Especially affected may be the highly attractive honeyeaters, birds that usually depend on the nectar produced by various eucalyptus and native shrubs of Australia. These days the flowering eucalypti of Australia hum with honeybee wing beats, as the insects cart away liquid refreshment that could conceivably sustain many more honeyeaters, honey possums (a miniature nectar-consuming marsupial), and the like. In addition, by occupying hollow trees, honeybee colonies may also be appropriating the homes of parrots and some small mammals that shelter in tree holes. Many of the magnificent parrots of Australia must find vacant hollow limbs if they are to nest, and yet these sites attract the colonial honeybees just as strongly.

Whether introduced European honeybees have caused ecological damage in North America or Australia remains a matter of debate. But one line of evidence suggests that honeybees can be powerful competitors for food and space. Enter the Africanized honeybees, a.k.a. the "killer" bees, which have recently displaced their own cousins, the European honeybees, from a wide swath of the Americas.

The first African bees were intentionally introduced into the Western Hemisphere in 1956 by a Brazilian beekeeper named Warwick Kerr, who hoped to cross African queen bees with the local Brazilian honeybees, which were derived from European ancestors. The idea was to create hybrids that would combine what were thought to be the best of both parental races, the docility of the European honeybee (which made them relatively easy for beekeepers to manage) and the supposedly higher honey production of the African bees. Although initially the introduced bees were kept in cages that were designed to prevent queens from escaping into the wild, within a year some had escaped, while others may have been removed prematurely from quarantine for distribution to local beekeepers.

The escaped colonies quickly gave rise to others in the standard honeybee fashion. During the season when pollen and nectar are most available, a colony produces new brood in the hexagonal cells composed of wax. Eggs

*Honeybee on desert
poppy*

laid by the resident queen become grubs which are fed pollen and concentrated flower nectar (honey) that have been stored in the hive. The grubs mature, pupate, and become new worker daughters, increasing the number of individuals capable of brood care, colony defense, and collection of pollen and nectar.

At the time of most rapid growth of the worker population, the queen will also lay some eggs destined to become drones and future queens. These individuals, unlike the workers whose ovaries are underdeveloped, are entirely capable of reproducing. After they have metamorphosed into adults, the new queens and drones fly from the colony to areas where others have assembled high in the air. Here drones pursue the females. Dozens or hundreds of males race after a passing queen, forming a comet of bees in the air. The male that catches a queen in midflight achieves a truly spectacular copulation, which ends shortly after the male explosively discharges his genitalia into the queen's "vagina." This event tears the male's abdomen apart and ends his life in a blaze of sexual excess. Thus, in the honeybee, unlike the praying mantis, there has been no prolonged debate about the occurrence of male sexual suicide. It happens.

Why it happens has not been resolved. Some persons have suggested that the combination of mucus and cuticle that the male fires into his partner blocks other males from copulating with her, and so helps the mating male monopolize the egg fertilizations that occur within the queen's body. Unhappily for this hypothesis, queens regularly mate several times during a nuptial flight; moreover, many queens are not content with just one nuptial excursion but take off on several different days for new rounds of mating. Thus, females typically copulate with a retinue of pursuers during each of several nuptial flights, always returning eventually to home base. Around this time, their mother, the established queen of the colony, takes off with roughly half the worker force to find a new home elsewhere. Her mated daughters remain behind, although in the days following departure of the queen some may also leave with a fraction of the worker force that did not leave when the old queen left. Each swarm eventually settles down in a hollow tree, rock crevice, or hole in the ground and the process of colony growth and fission repeats itself for as long as there are plenty of flowers about. When conditions are bad, the colony hunkers down and does the best it can with honey stored during times of plenty.

The original African escapees multiplied prodigiously in Brazil. A single colony of 10,000 to 20,000 Africanized bees can spin off as many as 64 descendant colonies in one year. As a result, an estimated 50 million to 100 million colonies of Africanized bees now call the New World home. A tidal wave of bees has surged through the Americas, some colonies going south into northern Argentina and others hurrying north across the Isthmus of Panama, through Mexico, and into the United States, which they have penetrated less than forty years after their initial "escape."

As the colonists advanced at a rate of several hundred kilometers a year, they utterly replaced their European counterparts, which were never numerous in tropical regions. Genetic analyses of the feral colonies of bees now present throughout Latin America show that their members are highly similar to the African, not European, race. The Africanized bees, which evolved in hot tropical Africa, apparently drive their European rivals to oblivion in hot tropical Latin America, proof of the potential for competitive displacement of their fellow insects by honeybees.

The spread of Africanized bees has been disastrous not just for the European race of honeybees in tropical America, but also for beekeepers in this part of the world. Africanized bees do not build large honey stores, but rather invest in the production of brood, which leads to the rapid fissioning of colonies. Beekeepers normally do everything they can to prevent swarming from occurring in their hives, especially by expanding the size of the hive in response to growing bee numbers so that crowding does not provide an impetus for colony splitting. The whole idea is to have a large worker force on hand during those periods of the year when nectar supplies are superabundant, so that the colony will generate a large honey store then. Africanized bees thwart this strategy because they tend to take off "prematurely" compared to European bees. Moreover, it is not easy for beekeepers to prevent swarming by manipulating hives that have been usurped by Africanized bees. These little devils are aggressive. Many beekeepers can handle European honeybees with bare hands and a minimum of protective gear. In contrast, Africanized bees often go berserk in defense of their colony, forcing their caretakers into elaborate head-to-toe outfits, lest they get covered with stings. Some beekeepers have had hundreds of enraged workers clinging to the veil of their protective headgear, squirting venom through the netting in their eagerness to sting something.

Once a beekeeper myself, I know firsthand that being stung provides a strong disincentive to go near one's hives. My uncle had a number of colonies at his place in New Jersey. With his help, I set up two hives on my parents' farmlet in southeastern Pennsylvania. We stocked one hive with a swarm of feral bees that we found in the woods near home. The bees were hanging in a mass from a low branch, surrounding the queen, while worker scouts hunted for a new nest cavity. Uncle Martin knocked the great lump of bees from the branch into a cardboard box. Then we carted them home and dumped them gently into a nice new hive with wax combs ready and waiting for them. The bees were probably thrilled; we certainly provided a far superior home than they could have found in a hollow tree near White Clay Creek. The transplants stayed put and began filling the combs with offspring, pollen, and nectar.

A beekeeper must keep an eye on his charges, adding extra supers to the hive, checking on disease, and removing the honeycombs at the appropriate season. I approached these tasks nervously, gloves on, hat and veil too, smoker in hand to puff smoke into the hive before I opened it. These precautions worked like a charm with the one colony of store-bought bees (purchased from an apiary supply house). Good European bees, these staggered about in the smoke and tolerated my inspection and behaved like gentlewomen. The feral bees were a different kettle of fish, to mix taxonomic metaphors. These harridans became quite upset with me and several found a way through my veil. Shortly thereafter they stung me.

When a honeybee worker stings, she leaves the barbed stinger and its associated poison sac in the skin of a vertebrate like me when she pulls away from the enemy. As a result, she soon dies of internal injuries, becoming an aggressive, as opposed to sexual, suicide. One consequence of the worker's ultimate sacrifice is that the stung one is punished more severely; the poison sac continues to empty even in the absence of its recent owner, providing the victim with a maximum dose of toxin. I had been told that, when stung by a honeybee, I should carefully scrape the detached sting apparatus out of my skin as quickly as possible, reducing the quantity of painful toxin injected into the body. In this maneuver, I was not to pinch the sting, which would supposedly propel the full dose of punishing chemicals into me. A recent paper in the *Lancet*, a prestigious medical journal, states that the mode of removal does not affect the amount of bee poison delivered by the sting. Now they tell me.

In order to carry out sting removal by any means, a calm and collected state of mind is recommended. I found it difficult to achieve the optimal mental state with several agitated bees buzzing about my head inside the veil. I kept wondering where I would get stung next. In addition, bee-keeper's gloves are not designed for delicate operations. The whole business left me unnerved and wary of bees. Appropriately, too, for my charges never missed an opportunity to seek me out whenever I came anywhere near the hives. To be zapped out of the blue for no good reason struck me as unfair. I gave up beekeeping not long after a mob of bees pursued me for several hundred feet down a hill and through a lilac hedge.

My feral European honeybees may have impressed me with their ferocity, but I can bless my lucky stars that they were not Africanized bees, which assault beekeepers with a persistence that can be overwhelming according to Mark Winston. Winston knows what he is talking about, having earned his Ph.D. in the study of Africanized honeybees in Latin America. It is the bees' aggressiveness that has attracted the most public attention and notoriety, as indicated by their killer bee moniker. The thought of dying from thousands of bee stings is no one's idea of a good way to go. While we were in Costa Rica in 1986, a botany student from the University of Miami happened to stumble on a colony on a precipitous mountainside. In trying to flee, he somehow became wedged in a rocky crevice. Unable to get away, he absorbed wave after wave of enraged bees, eventually dying from the cumulative effect of 8,000 stings.

Despite this widely reported horror story and a few others, Winston argues that the label "killer bee" is largely undeserved. For example, just two deaths per million can be attributed to stings of Africanized honeybees in Venezuela, where they are now fully established. On the other hand, the corresponding death rate from Europeanized bees in North America is 0.08 per million, with the deaths occurring because a few people are so severely allergic to bee toxin that a single sting can do them in. Thus, the presence of Africanized bees throughout North America might raise by 25-fold the death rate attributable to bees, not a trivial increase.

Luckily, not all of North America will be colonized by the Africanized bee, which cannot tolerate periods of heavy frost. A tropical-adapted species, it does best in hot, humid environments, not in cool, temperate zones where its competitor, the European honeybee, flourishes. The farther north one goes in the United States, the less likely people will become acquainted with the Africanized honeybee.

Those of us in Arizona, however, are destined to meet the bee, which has already made its initial forays across the border from Mexico. Someday soon the worker bees that cruise from flower to flower in my front yard are almost certainly going to be carrying African genes. In fact, neighbors up the street, who had long tolerated a feral colony of honeybees in their backyard, recently had to destroy it after a series of unprovoked attacks, which suggested that the site had been expropriated by an Africanized queen.

However, it is possible that the large population of European honeybees in Mexico and the United States will provide mates for many Africanized queens and drones, resulting in a higher degree of hybridization than occurred in tropical South America, where European bees have always been scarce. The greater the degree of hybridization between populations, the more likely Africanized bees are to become "Europeanized," with all the happy side effects associated with having more docile bees on the block.

Continuing to look on the sunny side of things, Mark Winston assures us that pollen- and nectar-gathering workers of Africanized bees are not especially aggressive. When the bees are swarming they also pose little risk to gardeners and the like. However, heaven help the person who happens to irritate an established colony. In their hive, workers have a perpetual chip on their collective shoulder, the better to protect the valuable brood and honey and wax that attract so many thieves in their native Africa.

However, the likelihood of nasty encounters with honeybees of any sort in North America has been greatly reduced in the past few years, thanks to yet another inadvertent introduction of a foreign species—a mite called *Varroa jacobsoni*. The mite once made a living by feeding on the brood of an Asian bee, *Apis cerana*, a close relative of our honeybee (*Apis mellifera*). The Asian bee copes reasonably well with the mite, having co-evolved with it over the eons. But the little parasite somehow transferred itself from its native host to nonnative honeybees about fifty years ago somewhere in Asia. Once having infected this new and largely defenseless host bee, the mite had a field day feasting on honeybee brood. Colonies infected by the mite rarely survive for long. Chinese beekeepers suffered huge losses to the parasite, which was declared persona non grata by beekeepers everywhere.

However, the worldwide trade in *Apis mellifera* led eventually to the inadvertent importation of some infested bees to the United States, probably from Brazil to Florida, according to Adrian Wenner and William Bushing. In 1987 several Floridian colonies shipped to Wisconsin died shortly after their arrival. The postmortem revealed that *Varroa* mites had established a beachhead in the United States. In just ten years, the mite spread from coast to coast, in large measure through interstate shipment of managed colonies by beekeepers. When declining colonies infested by the mites are assaulted and robbed of their honey and pollen stores by neighboring colonies, these once healthy hives inadvertently acquire deadly parasites along with their booty of food. Soon they too are dead, unless treated aggressively with a miticide.

The number of managed colonies in the United States has dropped by about one-third since the unfortunate introduction of *Varroa*, lowering honey production and, more important, reducing the ability of beekeepers to provide pollination services for fruit growers. Moreover, the once abundant feral colonies of honeybees have apparently been largely eliminated by the mite, since their hives cannot be treated with miticide by solicitous beekeepers. The era of free pollination services by feral bees has come to an end. How ironic that the introduced honeybee should itself fall victim to yet another alien species.

Agriculturalists worry night and day about the economic consequences of the decline in honeybee populations. They might even be willing to put up with Africanized bees, if these proved to be more resistant to the mite, a matter under current investigation. However, if it were possible to ship every last Africanized bee back to Africa, no doubt most of us would vote yes, and the sooner, the better. But at least some of the feared "killer" bees are probably here to stay, along with red brome, whiteflies, and a dreary list of other exotics, now expanded to include *Varroa* mites.

Admittedly, some alien species have wormed their way into my heart. One, a European gecko, appears to have colonized Arizona in the late 1950s or early 1960s, perhaps with an assist from the young children of a current colleague of mine, Wendell Minckley. His kids imported twelve geckos from Quatrocienagas, a town in northern Mexico where the lizard had been long established. The twelve imports were released around Minckley's house in Tempe, and from this beachhead, they appear to have skipped their way from neighborhood to neighborhood. At some point the species invaded our lot, a congenial event because *Hemidactylus tercicus* is a delightful little pink lizard, with the big splayed toes and bulging black eyes characteristic of geckos. The lizards spend the days lying under debris around the house and at night waggle their way up our walls to hang out by the entryway lights, waiting for an incoming brown moth or green lacewing. A quick dash and a snap of the mouth and the gecko has its evening meal. Late at night male geckos sometimes softly utter "geck-o," a vocalization designed to seduce females.

Attractive in the extreme, the alien gecko has yet another admirable quality, namely its failure to penetrate native habitats where it might compete with our desert lizards and other insect eaters. Instead, it is a true house gecko, utterly dependent on human habitation for its home and living. As a result, the Mediterranean gecko can remain on the front-yard list without causing discomfort to the ecologically correct.

I also wholly approve of many a cherished vegetable import: bok choy (from the Orient), Swiss chard (a native of Mediterranean Europe), various lettuces (derived from a wild lettuce from the Middle East), eggplant (from India), and a host of other naturalized immigrants whose mouth-watering images appear in my Burpee's catalogue. If I held my garden to the rules that govern plantings in my desert yard, I would be harvesting only corn, sweet potato, white potato, pumpkin, squash, tomato, and a few kinds of

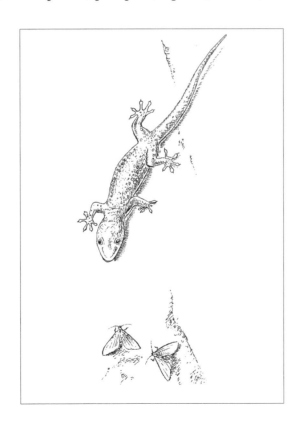

Mediterranean gecko

beans. I would not enjoy such a restriction, especially since I lack space for corn and pumpkins, and I doubt that sweet potato would flourish in Tempe.

And so, I waffle. Whiteflies may be the hobgoblin of little (and big) gardens and red brome is definitely the bane of my small front yard, but I am not going to reject every foreigner in a strange land, not if it means giving up my beloved Swiss chard and bok choy, edibles that I have come to appreciate for their forgiving attitude toward me as a gardener. After all, the business of acceptance works both ways.

TRANSIENTS

Mayflies are fragile insects of singularly elegant form and of very peculiar and interesting habits. . . . They are well known to all fishermen, and even philosophers have heard of them.

—J. G. NEEDHAM
The Biology of Mayflies

EVEN THOUGH I am neither philosopher nor fisherman, I too have heard of mayflies. Still I am deeply surprised to find a hundred or so of them flying in a swarm in the front yard this morning. Each individual in the mayfly assembly flies up a few feet, then flutters down, flies up, and flutters down, silvery wings glinting in the already penetrating sunlight, so that together the insects look like a fistful of self-propelled confetti. Mayflies. In Tempe. Far, far from the nearest water. On a hot summer's day in June. I have to look twice, three times, to be sure, but there is no doubt about it, especially after I secure my insect net and capture some members of the swarm. They are extremely thin and delicate-looking, almost anorexic, about three-quarters of an inch long, including three long hairs poking out from the abdomen tip. The mayflies appeal to me, except for those deep red-brown eyes, which bulge out of the head most bizarrely. If our eyes were proportionate to those of mayflies, they'd be as big as bowling balls.

I have every reason to be surprised, even astonished, by a swarm of big-eyed mayflies in our dry-as-a-bone, straight from Pioneer Rock and Gravel, Grade A imitation desert lawn. Mayflies are the classic *aquatic* insect, typically spending 99.9 percent of their lives underwater in the immature stage. The nymphs, as the subadult mayflies are called, lead lives of little glamour, creeping about on submerged rocks or vegetation, nibbling algae and getting picked off by fish. Trout, for example, hold mayfly nymphs in especially high regard. As a result, fly fishermen generally carry with them an array of ratty little lures, all grays and browns, looking like bundles of mouse fur, which sink when cast into trout streams, where, God willing, they may fool a benighted trout into thinking it is the real McCoy, a submerged nymph of mayfly species X, Y, or Z.

We have no trout streams in Tempe. A few recreational ponds, yes, encircled with paved sidewalks, concrete everywhere. Fishermen sit on the concrete and try to catch the occasional catfish unlucky enough to wander past. Arizona Game and Fish Department employees release flabby hatchery trout into the ponds during the coolest winter months to liven things up for our urban sportsmen.

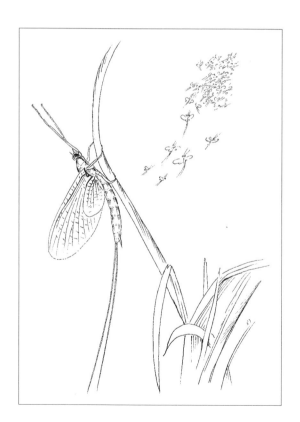

Male mayflies

I suppose some species of mayflies can make a go of it in Tempe's artificial ponds, or else there are mayflies in the cement-lined canals that carry drinking and irrigation water into town. Neither pond nor canal lies within a mile of our home. Nevertheless, adult mayflies have materialized right outside, from somewhere, to perform their mating dance. How unusual, since the mating swarms of typical mayflies occur over or very near water. The adults in these swarms have slipped out of their nymphal skin a day or two earlier, metamorphosing into a winged subadult (called a subimago) to rest quietly on rocks or plants near the water. The days when mayfly nymphs come out of the water to "hatch" into subimagoes are good days to be a trout fisherman out thrashing the stream with a fly line, provided that in one's arsenal of flies one has some dry (floating) flies as well as some wet ones, and among the dry flies a likeness of the appropriate subimago.

Even now the process of metamorphosis continues. One final molt, and the subimago becomes an adult with functional genitalia and the works. Typically, in the evening after this transformation, the now fully operational adult flies up to join others sailing about a few feet above the water or over a streamside clearing. In some species, almost the entire population of nymphs in a stream becomes adult on just three or four spring or summer days. The ability of so many mayflies to reach adulthood simultaneously impresses all who witness it: fishermen, naturalists, and a large and appreciative audience of trout.

The swarms of adults stay aloft for only a little while. Males use their bulging eyes to spot females flying into the swarm above them. They rush to grasp them with their long front legs, and they mate in the air before separating. The female then finds water that looks right to her and either showers it with eggs as she cruises over the stream or lands and floats on the water while depositing eggs, or she may even slip back underwater for a bout of submerged egg laying. Whether she succeeds in her endeavors depends on whether a trout interrupts the process by launching a sneak attack from below. Rising to the surface, a trout can inhale a floating mayfly with a sound like a person smacking his lips. When the mayflies mate, trout get fat—unless a dry-fly fisherman is on the stream casting the right mimic of the local adult mayfly. The choices are a bit overwhelming, including dark cahills, rusty spinners, sulfur duns, white-gloved howdies (as Dave Barry would say, "I am not making this up"), and about 150 other trout lures that resemble different mayflies.

Even those mayflies that copulate and lay their eggs successfully do not have long to celebrate their good fortune. The life span of most adult mayflies, male or female, is a few hours only, frenzied, orgiastic hours in which energy is expended as if there were no tomorrow, as indeed there is not. Exhausted females, having laid their clutch of eggs, soon die, their wings spread out on the water as their corpses float downstream. Exhausted males, energy spent to the last dime, flutter down into the stream to join their dead and dying compatriots.

Science can be poetic in its nomenclature. Mayflies have been assigned to an order called the Ephemeroptera, in honor of the brevity of their adult lives. Adult mayflies could not live longer by feeding, even if they wanted to, because they do not have functional jaws. The shortness of adulthood probably evolved because of predation by fish and birds, both groups so enthusiastic about mayfly hatches.

The argument goes like this. Adult female mayflies are very vulnerable to attack, particularly when laying eggs on the water's surface. Under these circumstances, the more individuals that metamorphose into adults

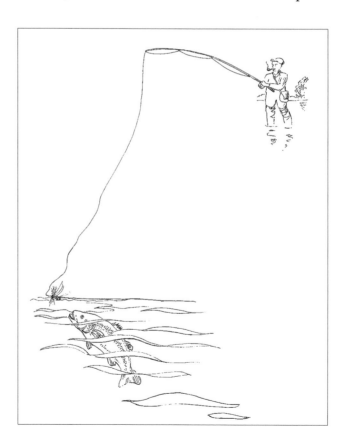

Fly-fishing

at the same time, the larger the population of prey, and the greater the chance that local predators will be overwhelmed by the number of edible targets. Those female mayflies that happen to come into adulthood on days of mass emergence will have a better shot at running the gauntlet and surviving long enough to lay their eggs than those that happen to become adults on a slow day, when predators quickly dispatch every last one. Thus predation favors females that happen to share emergence mechanisms that respond to the same cue, triggering the move to become adult on the same day.

If most females come out in a short period, males that happen to synchronize their metamorphosis with that of females will do better at surviving and mating than those that miss the Love Boat. If mating and egg-laying opportunities are limited to a few scattered days each year, then adults that have mouthparts and other devices to keep them alive for some time gain no reproductive advantage from investments in these structures. As a result, those adults that happen to put their all into living for a few hours during the peak times for mating and egg laying are more likely to leave descendants than those that hold back a little in order to extend their lives. If short-lived adults leave more offspring than longer-lived members of their species, soon ephemeral adults will become the rule and older individuals the exception.

Which makes the mayflies bouncing up and down in the air over my yard all the more amazing to me. They must have flown some distance to find this particular display ground. Moreover, any females that mate with the males fluttering up and down here will have to fly all the way back to water before laying their eggs, suggesting that these fragile insects possess sophisticated navigational abilities as well as energy reserves for a substantial round-trip without refueling.

I'm still impressed, even after reading a report by the mayfly specialist George Edmunds that some mayflies violate the "typical" pattern by mating up to a mile from water. In these species, adults gather far from their natal stream over landmarks, such as patches of bare ground or clearings in woods. Some mayflies in this group will happily swarm over a piece of dark plastic laid out on the ground. I have forbidden the use of plastic sheets in my front yard but, even so, it has some attribute that qualifies as a mayfly landmark, suitable for one of the species that travels long distances to find the right rendezvous site.

MY OCCASIONAL MAYFLIES are not the only transients, visitors, migrants, and nonresidents that add to the biological diversity of yard and garden, for which I am grateful. It is good to have the odd black-throated gray warbler or Lincoln's sparrow or green-tailed towhee slipping through on fall or spring migration adding spice to an avifauna dominated by a few ordinary permanent residents, namely mockingbirds, English sparrows, Inca doves, and house finches. In addition, since revolutionizing the front yard, I have visitors from neighboring desert regions that eschew lawns of Bermuda and nut grass. Curve-billed thrashers accommodate to urban landscapes after a fashion, but they are much more at home now that I have transformed the front yard. There they sprint from shrub to shrub, staying under cover unless they wish to probe the cowpies for termites. A much less frequent wanderer from the desert, a black-throated sparrow, once stopped for a few days to scamper about beneath the paloverde and brittlebushes. This bird, common in the true desert, usually rejects cityscapes, so I was flattered by its presence. Certain of my bird-watching colleagues and I like to brag in a cheerfully juvenile manner that we have added this species or that to *our* yard's bird list, when we get a new one. Without desert visitors and migrants, yard lists would be short and totally without conversational potential.

Although birds are by far the most familiar transients, there are some notable insects in this category as well and my revamped yard has attracted a few of these, among them the large milkweed bugs that I sometimes find on my milkweeds and at other times on my front-yard tomatoes. Let me digress for a moment to discuss why I have tomatoes available in the front yard for milkweed bugs to enjoy.

Growing tomatoes in Arizona is largely an exercise in masochism. Once temperatures get over 90–95°F, fruit will no longer set, which means that tomato production usually begins to fade by mid-April in Tempe. And by July, when weeks have passed since we had a high under 100°, not only are the plants unable to produce tomatoes, but those still on the vine are usually three-quarters cooked, with big patches of white sunburn on their upper surfaces. By then, the plants themselves are usually in terrible shape, leaves twisted and yellowed, terminal in both an aesthetic and a functional sense. After a quick mercy killing, out they go to the back alley to be dumped in a jumble to finish cooking in the shimmering heat.

So why do we bother, setting ourselves up for disappointment and dismay? Let me explain. I have just traveled to the local nursery to look over their sets. It is early February, winter in most parts of the country but a balmy day here, the sort of day that stimulates Phoenicians to phone their friends and relatives in Minnesota and Pennsylvania. I find it hard to conceive of the frozen soil and decaying mounds of soot-encrusted snow blanketing landscapes to the north; likewise, my gardener's mind rejects the possibility (indeed, the certainty) that Tempe will become a raging inferno in a few months. It is a season of optimism, a time to buy and a time to plant. I succumb to the season and purchase a set of six Yellow Pear tomatoes and a set of six Early Girl tomatoes. They are beautiful, green, and upright, not the yellowed, spindly, and rootbound specimens that we sometimes encounter here, and I must have them.

Never mind that there is no way that I can turn half my garden over to twelve tomato plants. Never mind that in the remaining weeks of February there is an excellent chance of frost that will destroy what I have purchased and will plant today. Never mind that, even if we are lucky and avoid a frost, my lovely young tomatoes will never grow into the gorgeous healthy shrubs covered with a bonanza of yellow and red fruits depicted on the tomato plant tag.

Tomatoes do that to people and have presumably been doing it ever since the ancient Mexicans decided to cultivate the ancestor of all today's

Black-throated sparrow

multitudinous tomato varieties, probably a cherry tomato of some sort. The Aztecs supposedly called their homegrown plant *xitomate*, giving us both the fruit and its name.

It is true that acceptance of the tomato in North America and Europe was not immediate. Because tomatoes belong to the Solanaceae, the nightshade family, which has some spectacularly poisonous members such as the deadly nightshade, the word got out that tomatoes were similarly toxic. The foliage of tomatoes does indeed contain alkaloids but they are not nearly so nasty as the alkaloids contained in certain other solanaceous plants, and in any event, the idea is not to eat the greenery of the tomato plant but its fruit. In North America, people grasped this concept between 1830 and 1840, driving the tomato to wild popularity. More than sixty varieties were available for farmers by 1880 and tomato producers have never looked back. Today Americans consume more pounds of tomatoes per annum than any other fruit or vegetable except potatoes. One somewhat dated source claims that the average person eats well over twenty pounds of processed tomatoes in a year, no doubt much of it in the form of ketchup.

When it comes to unprocessed tomatoes, however, disenchantment has grown in recent years with the commercially available varieties, which have been selected primarily for their ability to withstand the trials of picking, packaging, and travel to the supermarket, and not at all for their taste. As a result, these tomatoes hold up beautifully on the way to the store, a virtue that consumers find hard to appreciate in a salad, which is supposed to be eaten, not squeezed, hammered, or shipped to Tulsa or Boston.

Which brings us back to Early Girl and Yellow Pear tomatoes. These varieties may not be everyone's first choice but they are definitely edible, particularly in comparison with the long-distance travelers available in the supermarket. Since my Early Girls are just a few feet from the front yard to the kitchen, I can forgo durability in favor of palatability. I do have to compromise to some extent, selecting varieties that hurry through the stages from flowering to mature fruit, so that I can have some homegrown amendments to my salads before July reduces my plants to alley discards. The urgent necessity of western desert gardeners to get tomatoes to the table makes East Coast gardeners seem a bit frivolous when they compete to harvest the first ripe tomato of the season in their benign neighborhoods.

When the Early Girls do get around to turning red, after what seems an unreasonable time in the green holding pattern, I am not the only one to celebrate. The local English sparrows have learned that tomatoes are miniature oases. Once the fruit's outer skin is pierced, the sparrows viciously attack any tomatoes that I fail to protect with netting.

Although bird nets do a good job against the sparrows, the mesh is too coarse to screen out another smaller and admittedly less troublesome pest, the large milkweed bug, *Oncopeltus fasciatus*, another species that entomologists accept as a true bug. This bug has tube-like mouthparts, which it inserts into my tomatoes, possibly draining them of precious bodily fluids, although far, far more slowly than the local English sparrows. In fact, the bug may only be trying to feed on the seeds within the tomatoes, since milkweed seeds are its primary food. The creepies I find on my Early Girls may have moved to the garden from the two desert milkweeds in the yard. The big question: Where were they before taking up residence on my milkweeds and tomatoes?

Which leads us back to the topic of migration. It is entirely possible that my summer milkweed bugs are on migration from Mexico. Milkweed bugs are among the champion long-distance travelers in the insect world. Huge numbers drift along in diffuse flight heading north in the spring and summer, followed by retreats to the south in the fall. These flights are regulated by elaborate internal machinery that integrates information about the individual's age, the quality of the seeds it has been consuming, air temperatures, and day length. If, for example, food quality is falling, the bug's internal sensors produce signals that ultimately translate into declining levels of a chemical called juvenile hormone. Decreases in juvenile hormone in the blood in turn somehow make the bug restless and eventually cause it to take off for parts unknown. In effect, its migratory flight removes it from an area of declining food supplies and hustles it along to a lusher region elsewhere.

If flying migrants happen to settle in my front yard in early summer, the newly arrived bugs will find my milkweeds waiting with enough untapped resources to make colonization of these plants worth a try. The bugs that do settle down will have plenty of food to go along with high temperatures, long days, and short nights, the perfect combination to activate juvenile hormone production by the corpora allata, organs that control the output of this key chemical. With high levels of juvenile hormone coursing through the veins of the pioneers in the front yard, the bugs lose interest in flight and turn their attention to reproduction.

Their reproductive oddities are at least as interesting as their migratory abilities. For one thing, when a male milkweed bug sidles up to a female and persuades her to mate, which he does with a minimum of overt courtship, the couple often stay together *in copula* for hours and hours. Not only do most matings last extraordinarily long, but as soon as a couple has completed their time together and separated, male and female almost immediately pair off again with someone else. As a result of their enthusiasm for copulation, most of the bugs I find on my tomatoes are literally engaged, with one individual facing directly from the other, tied at the tips of their abdomens via their elaborately interlocked genitalia. When I disturb them, one bug takes the lead in scuttling to safety, while the other drags along behind, a dead weight on his or her partner.

The reduced mobility of mating pairs makes one wonder about the adaptive value of prolonged and frequent copulation by milkweed bugs. If the point of copulation from the male perspective is to transfer sufficient sperm to fertilize most or all of a partner's eggs, and if the point of copulation from the female perspective is to receive enough sperm to fertilize most or all of her eggs, then why not get the job done quickly and efficiently? And so it is in most insects, whose males can transfer quantities of sperm to their mates in short order, sometimes even in seconds. Moreover, the typical female insect receives and stores all the sperm she needs for a lifetime in one or a modest number of widely spaced copulations, thanks to her ability to stuff sperm away in her spermatheca, from which organ she will later retrieve stored male gametes to fertilize her eggs just prior to laying them.

Milkweed bug females have spermathecae and they receive enough sperm from just one male partner to fertilize all the eggs they will lay over the succeeding three weeks. Yet they almost never say no to a copulation and as a result spend most of their adult lives coupled to a male. There are a number of possible reasons why. For one thing, milkweed bugs are known to contain toxic glycosides. Some of these toxins are derived from their food, a topic we will take up later. The chemical cocktail contained in a milkweed bug makes it taste bad to praying mantises and other predators as well. When a praying mantis or a human being grasps the body of a milkweed bug, it oozes fluid droplets from special tissues along the upper edge of its body. These bubbles contain 500 times the glycoside concentration found in other parts of the insect. A single mouthful of glycoside exudate is usually enough to cause a bird to gag or a praying mantis to drop

*Copulating
milkweed bugs*

the bug. Like most bad-tasting, poisonous prey, milkweed bugs appear to advertise their unpalatability with bright colors, in this case a lively combination of black and red. The effectiveness of the advertisement may be enhanced when two milkweed bugs are coupled together, creating a larger, brighter warning to predators. No tasty meal here, they seem to be saying. If so, by mating at length, both male and female milkweed bugs might gain the same thing, improved protection from predators.

On the other hand, the benefits from leisurely copulations could be different for males and females. Males might extend copulation in order to prevent their partners from finding new mates and accepting sperm that would compete with their own; in other words, prolonged copulation may be a variation on the mate-guarding theme we first explored with the compost-inhabiting rove beetles that live on my parents' farm. Females might go along, copulating for hours on end, in order to free themselves from the sexual harassment that singletons would have to endure from courting males unwilling to take "no" for an answer.

Whatever the reason for their devotion to copulation, milkweed bugs are part of the yard's fauna, thanks to their ability to travel long distances and to track down suitable plants to feed upon while mating and mating again. They are not the only insect blessed with travel and searching skills. One famous migratory insect, the monarch butterfly, exhibits many behavioral parallels with our milkweed bugs. The monarch floats along beaches back East or rides the thermals in the Appalachians in spring and fall, heading north or south, in the appropriate season. Many are going to or coming from populations that take refuge during winter high in the mountains of central Mexico. Some will travel 2,000 miles to reach these sites, a fact established thanks to a wing-tagging project begun in the 1930s by the Canadian lepidopterist F. A. Urquhart.

Once in the transvolcanic belt to the west of Mexico City, the monarchs funnel upward to a handful of mountain ravines where literally millions of butterflies gather together to festoon the limbs and trunks of a select number of fir trees. Photographers swoon at the spectacle here in Mexico as well as along coastal California, where separate groups of monarchs spend winter months lethargically waiting for spring. Unfortunately, the highly localized wintering sites in Mexico are under siege from loggers, so much so that the East Coast populations of monarchs may actually be endangered, a thoroughly depressing prospect.

Monarchs truly migrate. The butterflies return to the places they have left in a regular back-and-forth pattern, although the Canadian monarch that made it to central Mexico will not itself fly all the way back north in the following spring. Instead it returns partway, and has some offspring, often somewhere along the Gulf Coast; these monarchs then take up the northward trek, reproducing in, say, Illinois, creating grandoffspring that may complete the journey to eastern Canada before some eventually turn around for the trip back to Mexico in the fall.

The monarch, sometimes called the wanderer, has not yet actually wandered into my yard, but I have on occasion had the pleasure of hosting a painted lady or buckeye butterfly. These visitors flutter into the yard, drift around for a bit, and dart away in short order, as if disappointed in my choice of landscape shrubs and weeds. Both the painted lady and the buckeye are renowned for their occasional outbursts of travel fever, creating mass flights of thousands, all headed in the same direction. Over several weeks one spring, I watched a seemingly endless parade of painted ladies crossing the flank of nearby Usery Mountain, every last one following the same west-northwesterly route, even though the butterflies were often flying into headwinds from the west. The ability of the butterfly to maintain a constant orientation in the face of opposing or variable winds suggests that it employs a standard navigational aid or two, such as sun position or the earth's magnetic field, to get wherever it is going.

Painted lady butterfly

Unlike migrating monarchs, traveling painted ladies and buckeyes rarely retrace their route the very next season (nor do their offspring). At least in the western United States, the great butterfly flights probably occur when conditions for population growth deteriorate in one area, leading eventually to a mass exodus of adults in search of a better situation, though the place may be far away and hard to find. Just why all of the butterflies should choose a similar flight path has not been explained as far as I can tell. Nevertheless, one minor consequence of the readiness of painted ladies and some other butterflies to get up and go is that yards like mine are likely to receive them as transients now and then, and the yard is the richer for it. *Sic transit gloria.*

PARTHENOGENS AND
POISON EATERS

The Desert Ridge master-planned community has become a bonanza for
Valley home builders. The 5,700 acre development, off Tatum Boulevard
north of the Central Arizona Project canal, features easy access to
shopping and business areas. Plans call for eight-story office buildings,
one- to four-story garden offices, a power center, a resort, two 18-hole
golf courses and 25,000 homes.

—*Arizona Republic,* January 13, 1996

OUR AGED HOUSE occupies the slot for Unit Six of the
Cavalier Campus development, no doubt so named
because it sits less than two miles from the campus
of Arizona State University. Perhaps the developers
hoped to attract university employees as purchasers. In this they succeeded, although the original owners of our home were not academics.
When they pulled up stakes for a more recent development, we inherited a
halfhearted flower border in the backyard with a dozen dejected roses,
which did not thrive under new management. An erratic watering schedule and too much shade combined to sentence the roses to a lingering
death. Before they expired, few roses bloomed, since most of the buds
failed to open, but rather drooped and withered instead.

Aphids were to blame. Ere a rose started to flower, the bugs appeared
from nowhere in massive numbers in a sort of Woodstock for insects. Traf-

fic jams formed on the stems of developing buds, which shortly wilted, blackened, and died.

Offended by these events, I contemplated several options, ultimately deciding to go with a systemic insecticide that I cast around the roses, watering the stuff deeply into the soil as instructed. These ministrations did no good. The rosebushes eventually followed other small vegetative disasters out to the alley, where they lay in a forlorn heap until swept away by the trashman. Years later I took my still largely full yellow-and-red cylinder of poison to the Tempe dump on a day when the authorities were accepting for disposal the full panoply of industrial-strength toxins available to civilized man. I was glad to have it out of the tool shed.

Now that I have brittlebush, penstemons, and milkweeds in place of roses, I still have aphids, but they appear to coexist much more comfortably with my desert-adapted plants than with garden roses out of their temperate-zone element. Often a solid tube of greenish aphids covers the stem of a brittlebush flowerbud without killing the flower, which opens up its cheerful yellow face more or less on schedule. I have no doubt that aphids exact some toll from the brittlebush they infect, because they draw off nutrients that the plant manufactured for its own use. The considerable quantity of sugary fluid thus diverted into the collective gut of the thousands of aphids feasting on one of my brittlebushes could have helped the plant support more flowers or a more prolific root system or new leaf tissue.

Still, at least superficially, my brittlebush get along, perhaps because conditions for their growth are so good in my front yard that they can shrug off the assaults of aphid battalions. In nature, brittlebush have to compete fiercely with relatively closely packed neighbors for limited soil water. I have relieved my brittlebush of this burden by keeping them well apart, enabling them to sprawl across the gravel uninhibited, as nature would never allow.

The pale green aphids on my pampered brittlebushes belong to one of the hundreds of species placed in the genus *Uroleucon*. Almost all of these aphids specialize in making life somewhat more difficult for members of the aster family (to which brittlebush belongs, as anyone can see when it produces its daisy-like flowers). Other plant families have their own quite different aphids, which restrict themselves to their special hosts. So, for example, my front-yard milkweeds are colonized by *Aphis nerii*. Their common name, the oleander aphid, is something of a misnomer, given their strong preference for milkweeds in the family Asclepiadaceae, although they will settle for oleander if this ornamental shrub (which belongs to the family Apocynaceae) is all that is available.

The specialization in host plant selection by various aphid species is an intriguing business. Why should one species love only brittlebush, another desert milkweed, with still others tied to their own special plant victims? The issue is not yet resolved, despite much thought on the matter by assorted ecologists. Many have suggested that trade-offs exist along the spectrum from specialized to generalized consumers. The specialist presumably can afford to have its feeding and digestion devices very tightly designed for its special host. The result should be extremely high efficiency in securing and processing the chemical nutrients present in that plant species. The more generalized consumer must have physiological systems versatile enough to deal with a wider range of plants, which should mean that they won't cope as well with any one host. A jack-of-all-trades, but master of none. On the other hand, the generalist depends less on a single source of food, and so may thrive while specialists suffer when their special food declines or disappears entirely.

While the trade-off theory seems logical enough, relatively few tests of the idea have been attempted, particularly with aphids. One such study did examine two species of *Uroleucon*, close relatives of the brittlebush aphid. The two observed species are both found on the same host plant in

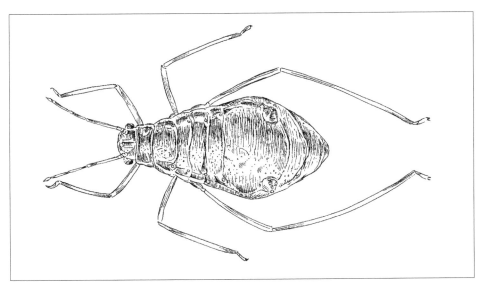

Marvelous aphids

Michigan, one being limited to a single goldenrod, the other more catholic in its diet, accepting goldenrod and other plant species as well. If specialization really does foster feeding efficiency, then the extreme goldenrod specialist should develop faster, grow larger, and be more fecund on this food source than the generalist. And so it is, but only for a few weeks in the spring. By the time the summer comes round, the generalist aphid is every bit the match for its cousin in terms of growth and reproduction. The equivalence of specialist and generalist carries on right into the fall. These results offer little encouragement for the hypothesis that the goldenrod specialist aphid is superefficient at turning goldenrod nutrients into aphid bodies and aphid babies.

Whether specialized or generalized, aphids are clearly well designed to feast upon whatever they prefer as hosts. Perhaps too nicely designed for the good of some hosts, or so I fear with respect to the two rather stunted desert milkweeds in my front yard. Year after year rolls by and yet my milkweeds seem little larger than when I transplanted them from their nursery pots. To help them grow, I even run the garden hose out to them on occasion.

Despite the extra water, my milkweeds still seem sickly, perhaps because they suffer more from aphids than do the exuberant brittlebush. Although I have never tried my hand at controlling the brittlebush aphid, I confess to aphidicide when it comes to the milkweed-oleander aphid. My biocontrol project involves running my fingers up and down their stems during times when the plants are overrun with bright yellow *Aphis nerii*. After an aphid squash, my fingers come away sticky and stained with the cadmium yellow of a hundred aphid bodies.

Sometimes I also get a bit more technological in the aphid war when I turn on the insects with a sprayer filled with mildly soapy water. The spray knocks them from their perches while coating them with a suffocating film of soap.

Despite my murderous impulses, however, I have never been able to rid my milkweeds of hungry aphids. That fraction of the population that escapes being crushed to death or soaped out of existence quickly begins to reproduce. In this, milkweed aphids are remarkably adept because they, like all other aphids, can engage in asexual reproduction, parthenogenesis, virgin births. In fact, the population of plump little yellow bugs on my milkweed stems is entirely female, not a male in sight.

Each female produces eggs that are not fertilized by sperm, and yet despite this apparent handicap, they develop perfectly well inside their mother's body. In due course a minute but well-formed infant slowly exits her mother's birth canal at the tip of her abdomen. The miniature aphid waves her legs about, contacts the stem, rights herself, dries her wings and body cuticle, and then promptly stabs her fluid-receiving proboscis into the plant. Standing side by side, mother and daughter extract the material wherewithal to live on and to reproduce again. Soon there are four aphids, where once perched two, shortly to become eight and then sixteen, each capable of having still more daughters with a mere ten days between generations.

The ability of aphids to forgo the production of sons means that a population of asexual aphids has twice the growth potential of a population of sexual individuals (in which sons and daughters are produced with equal frequency, which is the typical recipe for a sexual species). Since male animals usually contribute nothing other than their sperm toward the production of offspring, the number of females generally determines the number of offspring in the next generation. In a sexually reproducing population of 100 aphids, there would be only about 50 females capable of creating that next generation. In an asexual population of 100 parthenogens, all will be females that contribute directly to the next cohort.

Even so, not all aphids eschew sex. In some species, one special generation of females produces both sons and daughters. Unlike their mothers and grandmothers before them, which only made one kind of egg, each with an exact duplicate of the female's genetic information, these special females generate two kinds of eggs that differ in the number of X chromosomes that they contain. (The X or sex chromosome is one of several different chromosomes that aphids possess; each is typically present in two copies; thus, there are usually two X chromosomes, two copies of chromosome 1, two of chromosome 2, and so on.)

If an egg produced by the special female has the complete double complement of chromosomes, including two X chromosomes, it will develop into a daughter capable of sexual reproduction. If, on the other hand, the egg has a single X chromosome but two copies of all the other chromosomes, it will become a sperm-producing (X0) male.

Once they achieve adulthood, X0 males and XX females get to work. The females of this generation manufacture distinctive eggs that have only one copy, not two, of all the chromosomes, including the sex chromosome. The males produce sperm, which also bear just one complement of chromosomes, but with the specification that every sperm made contains an X chromosome. (In the process of manufacturing sperm, those cells that are formed without an X chromosome are simply eliminated.) Thus, when male and female aphids consummate their union, the fertilized eggs will carry a complete double set of chromosomes, including two X chromosomes. All of which means that father aphids can never have X0 sons, only XX daughters.

This new generation of daughters has nothing to do with sex. Instead they reproduce parthenogenetically in the manner of their grandmothers and great-grandmothers, making only XX eggs which become daughters that do the same thing, and so on. Eventually the cycle comes round again to the generation in which females parthenogenetically bear both sons and daughters, setting the stage for one more generation of aphids capable of copulation and sexual reproduction.

The whole business seems more elaborate than it has to be, and I have not employed the full dictionary of jargon associated with aphid life history, which includes titles for various types of aphids: fundatrices, fundatrigeniae, sexuparae, exules, and the like. Nor have I commented on other complexities in the story, such as the occurrence of both winged and wingless parthenogenetic females in some species. Up to eight different forms can be produced within a single clone. Must nature be so complicated? Why don't female aphids simply churn out daughters that are duplicates of themselves in every way?

Declining food supplies provide a reason why female aphids might produce winged daughters under some circumstances. For a time, mothers, daughters, and granddaughters have plenty of food in the plant they are mutually consuming. Females that have babies genetically and physically identical to them do fine. But all good things come to an end. As plants mature and seasons change, the nutritional quality and quantity of the fluids available to feeding aphids often drop sharply. Time to move on. Now females that turn out winged daughters give these offspring the ability to escape a deteriorating situation in order to maintain the lineage elsewhere.

Other factors may favor females able to produce sons and daughters for the single sexual generation in the annual cycle of some aphids. As an aphid population grows asexually, genetic differences between individuals are low. In fact, they may be absent altogether, if all the females on a plant descended clonally from a single foundress. Suppose, then, a tiny aphid-destroying virus or bacterium comes along. If the parasite infects one aphid, its offspring will run rampant, surrounded as they are by an abundance of suitable hosts with the same genetic makeup, and thus the same vulnerability to the parasite. Clones of genetically identical aphids will wither away in the face of this kind of onslaught.

One way to counter such a calamity is to reshuffle the genetic deck from time to time, and this happens when female aphids produce males and females that mate, bringing together novel combinations of genes as genetically distinctive sperm from one aphid lineage combine with genetically distinctive eggs from another lineage. Some of the resulting offspring may be lucky enough to put up new genetic barriers to infection. Thus, by occasionally making males and having sex, different aphid lineages exchange genes and may sidestep certain parasites.

The yellow milkweed aphid is interesting in this regard because its use of the sexual option varies over the current range of the species. The aphid may have originated in eastern Asia, only to have been introduced fairly recently into North and Central America. In Japan the species has a sexual generation every year, possibly to help lineages escape the parasites endemic to the region in which the milkweed aphid has evolved. In contrast, the milkweed aphid now in the Western Hemisphere reproduces strictly asexually, possibly because it has escaped the enemies that make sexual reproduction a worthwhile investment in its original homeland.

I do not know, however, whether the yellow aphids drinking from my milkweeds are freer from parasites than their Japanese cousins. But the aphid is certainly fecund here. Soon after the first colonist appears in the early spring and slips her beak into the milkweed, daughters, grand-daughters, and great-granddaughters *ad infinitum* make life miserable for the plant as milkweed calories flow into aphid offspring.

Given the seasonal abundance of bright yellow milkweed aphids, one wonders why they do not attract more attention from vertebrate enemies other than the yard attendant at his house on Loyola Drive. Some potential aphid eaters restlessly patrol the area. For example, verdins, the desert relative of chickadees, often hunt carefully through my brittlebushes. I have also seen them in the garden proper inspecting the undersides of squash leaves. There they almost certainly find and eat the powdery gray aphid species so expert at sucking squash fluids. I applaud the verdin's activities on behalf of my squash, but have never seen them working over milkweed aphids.

Perhaps they do when I am not looking, or perhaps milkweed aphids constitute noxious inedibilia from a verdin's perspective. The possibility that they are inedible is a strong one, but to understand why we must first digress to examine how milkweeds defend themselves against most of their would-be consumers. The milkweed out front belongs to a genus, *Asclepias*, noted for having two lines of chemical defense. First, there is the milky sap that gives the plant its common name. This material is stored under pressure in a network of canals throughout leaves and stems. Injure a leaf or stem, and the stuff oozes out. Exposure to the air quickly converts the latex into glue that will gum up the jaws of a plant-feeding insect. Not surprisingly, most small plant eaters give milkweeds a wide berth, simply because the sticky sap makes it impossible for them to eat the plant. But this gluey barrier has been overcome by some special consumers. For example, certain caterpillars cope with latex flows by snipping a main canal low on the midrib of a leaf before turning their attention to leaf material above the wound. The open canal releases milky latex and lowers the pressure in the system, so that the fluid cannot come welling out of the network of canals farther out on the leaf.

Interestingly, variations on this tactic have evolved in other insects that deal with different kinds of latex-producing plants. Perhaps the most familiar "milkweed" that is not an *Asclepias* is lettuce, *Lactuca sativa*, a member of the Compositae. Although approximately 6,500 years of domestication have greatly lowered the latex content of the plant, it can still release a sticky exudate from wounds on leaves and flowers. Just by *walking* on these tar-baby plants, an aphid or whitefly may acquire more and more latex on its feet, eventually immobilizing the little pests.

For every problem, however, there is a counteradaptation. The sticky trap defense of lettuce has been overcome by the cabbage looper caterpillar, a common garden pest that attacks many more vegetables than cabbage. The looper gnaws a trench across the entire base of a lettuce leaf, or a prominent lobe of same, interrupting the diffuse network of canals that some members of the genus *Lactuca* employ. Bubbles of latex form along the trench, but after this operation, the distal portion of the leaf can be consumed with impunity by the wily caterpillar.

The latex of lettuces and milkweeds offers a gummy mechanical obstacle to leaf eaters, some of whom reply with their vein-cutting or trench-gnawing tricks. But if a gummy barrier fails to deter, *Asclepias* milkweeds bring on a second line of defense—systemic poisons, the cardiac glycosides in their tissues. This poison or an allied chemical tastes extremely unpleasant or flat-out kills most herbivores. So, touché. But wait, just as some insects have overcome the milky glue defense, so too have others evolved special digestive systems equipped with the necessary enzymes and other chemical machinery to inactivate or sequester the plant's poisons. For example, several African grasshoppers not only feast comfortably on various members of the Asclepiadaceae, but also almost certainly incorporate the poisons in their food plants in their bodies for use against their own consumers. The smell of one milkweed-eating grasshopper literally nauseates humans that come within a few meters; presumably the hopper's potential predators also stay well clear of this mobile stinkbomb.

Poison recycling also occurs in some well-studied North American milk-

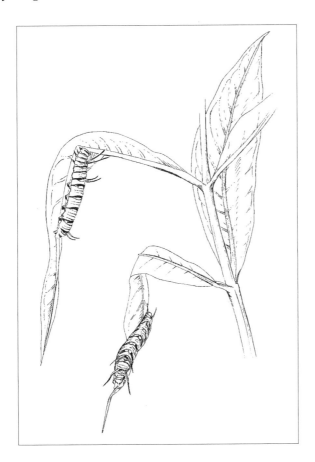

*Queen butterfly
larvae feeding on
milkweed*

weed eaters. Here in the desert, I often find the larvae of queen butterflies growing fat on milkweeds by roadsides or desert washes. The butterflies have evolved the biochemical wherewithal to ingest milkweed toxins safely and to store the poisons in their tissues for personal defense. A bird that tries to eat a queen butterfly larva or adult gets a mouthful of evil-tasting milkweed substances that usually induces it to release its prey pronto. Should the bird insist on swallowing its victim, it throws up a short time later, voiding the dangerous chemicals in its recent meal. The experience is evidently memorable. Captive birds that have vomited after feeding on one toxic queen or monarch butterfly (another milkweed specialist) almost never come back for seconds.

Which brings us back to milkweed aphids. If these bugs also incorporate the plant's toxins in their bodies, they would be unappetizing snacks for verdins and the like. Ladybird beetles reputedly find these aphids unacceptable, suggesting that the species is poisonous. In this light, their bright color might act as an advertisement of unpalatability, the better to help some predators learn from experience to leave them alone. As noted, chemically protected prey usually flash bright colors, often yellows and reds, sometimes intermixed with bands of black. Apparently most birds and other predators can more easily remember the negative consequences of handling prey with these color patterns as opposed to drabber color combos.

I suppose I could test the unpalatability hypothesis by sampling milkweed aphids and brittlebush aphids. I expect that the former would taste much worse than the latter. However, I have yet to carry out the experiment, but not because of an aversion to consuming insects. Although in our society eating bugs is taboo, I am not so squeamish nor are people living in many other places. A friend of mine, now working as an entomologist in Cambodia, tells me that the Cambodians snack on a wide range of insects. Admittedly, the enthusiasm of these people for insects is partly the legacy of living under the viciously cruel Khmer Rouge during the 1970s, when a starving people ate whatever they could find. My friend is not food deprived, but even so, he reports that mole crickets fried in oil and garlic are both crunchy and tasty. My preferences run toward uncooked insects, having discovered that one can derive considerable notoriety by demonstrating the edibility of some bugs at parties and the like. Although publicly downing a mealworm or Indian house cricket may have helped me consolidate a certain reputation, I am happy to report that my digestive tract has not suffered when dealing with these hors d'oeuvres.

But if it were true that milkweed aphids contain repellent cardiac glycosides, eating even a few of them might be a very bad idea. Humans are highly sensitive to these digitalis-like substances because they interfere with normal functioning of heart muscle in our species. No wonder the lepidopterist Lincoln Brower cautions against eating even one monarch butterfly that has been fed as a larva on one of the more poisonous milkweed species. Most people require little instruction on this point. Thus, if I were to experience cardiac arrest after eating a handful of milkweed aphids, I would provide powerful evidence in favor of the idea that these insects are chemically protected prey. However, I shall forgo the gratification that might come from a definitive test of this hypothesis.

AT LEAST ONE predator of milkweed aphids can consume them even if they are laced with cardiac glycosides. On one of my desert milkweeds I found a small pale mottled grub four or five times larger than the yellow milkweed aphids perched on the stem nearby. This highly nondescript little worm, the larva of a syrphid (hover) fly, had grasped an aphid with its mouthparts and was in the process of draining its victim. In trying to photograph the predatory grublet, I touched it, causing it to writhe about and then drop from the waxy green stem. The larva fell a couple of feet to the ground, where I lost it in the gravel beneath the milkweed. Early the next morning, however, when I checked the plant again, I found a syrphid larva partway up a long stalk of milkweed. The creature had wrapped its body around the smooth stem and was inching upward in a way that reminded me of a human trying to climb a flagpole. A cluster of aphids fed unconcernedly at the top of the stem, unaware of what was heading their way.

Adult syrphid flies are conspicuous and widespread dipterans. The males of many species spend hours hovering in shafts of sunlight, waiting for females to fly to them for mating, thus the common name "hover fly." They are, according to the entomologist Harold Oldroyd, "almost the only family of flies that everyone knows, and everyone likes." I question whether Oldroyd has this right since in my experience most folks have only the sketchiest knowledge of fly families and are not particularly enthusiastic about any of them. However, it is true that syrphids as a group are harm-

*Syrphid fly adults
and larva feeding
on aphids*

less, inoffensive, uninterested in humans, and delicately attractive. A good many resemble small wasps and bees, a deception the flies use to fool predatory birds that have been stung by a real thing.

The larvae of some syrphids feed on detritus and decaying plant matter, but others, as noted above, are specialists at dispatching aphids. It is not uncommon to observe a grub mowing down one prey after the other, while the doomed next in line sips plant juices oblivious to its fate.

Actually, some aphids do not go gently into that good night, but fight back when a predatory syrphid or some other arthropod enemy, such as a ladybird beetle larva, makes a visit to their clone. The first person to make this discovery was Shigeyuki Aoki, who identified aphid "soldiers" in 1977 while studying a Taiwanese species, *Colophina clematis*. Since the members of a single species of aphid often come in a kaleidoscope of forms, small and large, winged and wingless, plump and skinny, Aoki would not have been surprised to find diverse and distinctive forms in "his" aphid. But what caught his attention was the powerful piercing beak of certain members of the clones he inspected. These special individuals had sword-like mouthparts that were thicker than those of their genetically identical sisters. Furthermore, their front legs were also much more robust and powerful than those of their clonemates.

Aoki found that these well-armed individuals used their stout legs to grasp syrphid larvae firmly, after which they inserted their stabbing mouthparts into their enemies. When they pierced the skin of the larva, it often bled to death. No tears were shed by the killer aphids.

Taiwanese aphids are not the only ones blessed with kamikaze soldiers. In fact, the soldier caste has evolved in at least five different groups of aphids around the world, including some European and North American *Pemphigus*. These aphids live in hollow plant galls and are protected by enforcers with piercing mouthparts and thickened *hindlegs*, which they use when a syrphid larva attempts to enter the opening into a home gall in search of prey. Many soldiers die trying to puncture the predator's skin with their mouthparts and spiny hindlegs (on average it required about twenty aphid martyrs to kill one syrphid grub in William Foster's experiments with these insects). The importance of soldiers became evident when Foster removed all the defenders from certain groups, after which the unmolested syrphid larva ate all the remaining aphids.

Here in Arizona, Nancy Moran has also found a species of *Pemphigus* that has soldiers. These aphids live in galls on cottonwood poplar leaves. (Alas, a giant cottonwood would overpower my front yard; otherwise I might have the thrill of watching *Pemphigus* soldiers just by stepping out the front door.) When a foundress female has young, they are born with armored cuticle and powerful legs, as well as a willingness to go on patrol outside the small opening to their gall. Should they encounter a predatory fly grub or lacewing larva near the entrance to their home, they attack in number, grabbing and stabbing. Not surprisingly, grubs so treated writhe about and drop off the gall, carrying their attackers to their doom. However, the remaining sisters in the gall no longer have to contend with a deadly intruder in their midst. As the surviving aphids mature, they go through a series of molts and lose the ability to behave like soldiers. Instead they rely on their younger sisters as these are produced. If all goes well, some aphids will reach maturity and fly away from the gall to reproduce elsewhere, having secured this opportunity thanks in part to the sacrifice of deceased sisters.

The willingness of soldier aphids, whether in Europe, Arizona, or Taiwan, to launch suicidal attacks on a predatory fly larva or other similar enemy of aphids is most impressive—and most unusual. Truly suicidal

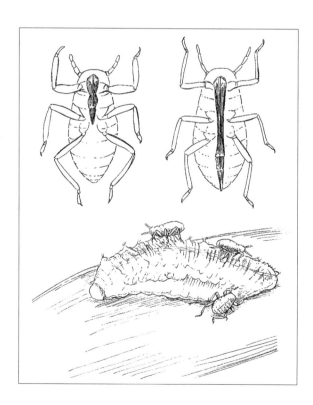

*A fly larva under
attack from soldier
aphids (the form on
the left)*

self-sacrificing behavior has evolved in only a handful of insects other than these few aphids, notably those social bees, wasps, ants, and termites whose sterile or reproductively inactive workers sometimes throw their lives away. Why? An answer lies in kinship. The beneficiaries of altruistic acts by insects usually are close relatives of their "heroic" nestmates. If by giving her life away, a worker ant or wasp saves the lives of some of her sisters, she will have saved some of her genes as well. If her saved sisters go on to reproduce, then they will carry the "genes for altruism" into another generation.

The same kind of evolutionary explanation might solve the mystery of soldier suicide in aphids. Those individuals that gave up their lives in defense of others in a parthenogenetically produced clone are benefiting sisters that are genetically identical to them. You do not get any more closely related than that. So although the cost of a suicidal attack is obviously high for the soldier, the benefit can be great as well, provided some sisters live to reproduce as a result.

Given this way of looking at things, the real puzzle becomes why sterile aphid soldiers are limited to just 50 or so of the roughly 4,600 known species of aphids. Part of the answer to the puzzle lies in the recognition that aphid soldiers require energy and resources, but are unlikely to reproduce and pass on their mother's genes. Instead, they must "justify their existence" by helping their fertile clonemates survive. Perhaps they can have this effect only under highly restrictive conditions that have yet to be fully identified by the small band of specialists working on aphid altruism.

And here we shall leave the matter. But let us, before moving on, ponder our attitude toward aphids. Pests they sometimes are, but as nuisances go, these creatures are fabulous. They reproduce with or without the curious beings we call males. They develop flexibly enough so that two members of the same clone may look like two entirely different species. Some have even evolved a soldier caste ready to make the ultimate sacrifice for their clone. For creatures so marvelous, the loss of a few roses or even the stunting of a native milkweed does not seem too great a sacrifice, as far as I am concerned.

PARADISE REGAINED

There is no season such delight can bring,
as summer, autumn, winter, and the spring.

—WILLIAM BROWNE

WILLIAM BROWNE DID not live in Phoenix, Arizona. Walking out the front door on my way to the garden, I plunge into a heat so physical that it threatens to grab me by the throat and throw me back into the house. The broiled gravel in the front yard seems to flinch like kernels about to explode in the popcorn popper. The concrete in the driveway is doing its Chernobyl imitation and will radiate heat for hours after sunset.

The garden has foundered on this summer reef of unbearable temperatures. The once cocky green leaves of the zucchinis now droop like closed parasols, just barely transpiring. In addition to the climatic horrors of an Arizonan summer midday, they struggle against a virus that has cast a silvery sheen over mature leaves while withering new buds and aborting pale green baby fruits. The asparagus beans cling to black plastic netting, more dead than alive. I no longer hold out hope for the Yellow Pear tomatoes, wrinkled and insipid with heat stroke. A late Early Girl turns a whitened cheek to the sun, a bit of remnant red flesh sheltered in its own shade.

Vegetable delights may be in short supply, the temperatures may be out of control, but I still view my front yard as a pocket paradise of sorts because I have not exhausted its harvest of insect discoveries. Eight years have passed since the Kubota tractor and I rearranged my property, but new bugs still keep coming. It may be July in the desert, but a host of insects don't mind, as the *Arizona Republic* informs its readers with a headline featuring the words "muggy" and "buggy." Underneath, an article broadcasts advice on how to dispatch the cockroach, cricket, ant, and company with well-aimed swats, insecticidal sprays, poison baits, and the like. The journalist even suggests vacuuming up certain special offenders. I, however, shall not terminate my summer insects. They are far more interesting alive than dead.

Take the harmless little *Cerceris* wasps I encountered three weeks ago on a mid-afternoon excursion to the garden. As I approached the two surviving Japanese eggplants, I noticed four or five wasps, only a half inch long but waspish nonetheless, zigzagging around the leaves of the plants. Peering as closely as possible, I realized that I had never seen this species in any previous summer. The flurry of activity caused by my arrival subsided gradually when I stopped moving. One by one, the wasps landed on a leaf edge and ducked underneath to perch in the shade. The decision to get out of the sun made sense to me, since it was 104 even in the shade while approximating hell elsewhere.

I stood my ground long enough to see the wasps make regular forays out into the sunshine, traveling round the eggplant. When one wasp met another in flight, each briefly acknowledged the other's presence with a swirling circle before continuing on. I noticed that a relatively large wasp seemed to come back to the same leaf several times in a row. Once a flier found and buzzed a perched companion before settling on the underside of the same leaf.

No great battle or other drama took place before I retreated into our air-conditioned house to lower my body temperature and collect my thoughts. But my mood, which had been in a glum summer funk, had improved. I had been offered a new puzzle to solve, a new species with its own gambits to decipher. I had some guesses about what was going on, having studied

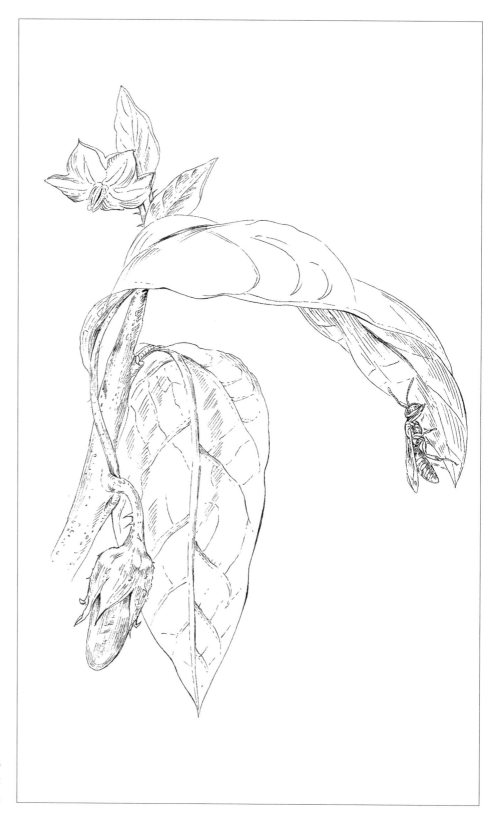

some close relatives of *Cerceris* wasps more than twenty years previously in the desert near Phoenix. Males of that wasp also gathered in groups on prominent bushes, with individuals having favorite perches used as launching pads for sallies out from and then back to their resting spots. Very rarely, females approached the cluster of males and selected one for a mate.

I guessed that my wasp assembly also consisted of a gang of males waiting for sexually receptive females to come to the eggplant for a tryst. How to check on my guesses? Step 1: Catch a wasp. I picked up a vial and headed outdoors again. The resident wasps responded to my approach as before and then settled back on their upside-down perches. I eased the vial underneath a *Cerceris*, only to have it nervously zip off before I could trap the creature. I tried again with another *Cerceris*. Same result.

Okay. So much for the vial. Time for a new technique. I found another perched wasp, and this time I slowly eased my thumb under the leaf while my forefinger moved in parallel above the leaf. Then I quickly brought thumb and forefinger together over the body of my prey. Voilà. I had him. Even if he had been a she, I wouldn't have been stung. This wasp was too small to do damage. In any case, I could tell I had a stinger-free male by the white patch on the wasp's face, a characteristic feature in a great many male wasps and bees (females typically having darker faces), and a useful one for those of us who like to know the gender of the wasps and bees we handle.

Step 2: Now that I had a male under control, time to mark him. I happened to have some water-based acrylic paints for just such an occasion. Back into the house I went, carefully gripping the wasp, not so firmly as to damage the little guy, but not so loosely as to have him escape indoors. Popping the top of a jar of Liquitex bright aqua green with my free hand, I picked up a fine tipped camel's-hair brush and dabbed it in the paint. Carefully, I transferred a bit of paint onto the black thorax of my captive. In painting bees and wasps, be careful not to apply the material to the insects' wings, particularly when working with small species, if you wish to have the subject fly away when released. Green Dot obligingly did fly after I returned him to the eggplant, cruising over to perch on a nearby zucchini leaf. There he groomed himself fastidiously, removing some of the scents and oils that he had acquired from my fingertips. Then he returned, unimpaired, to patrol the area and perch again under an eggplant leaf.

Step 3: I repeated the procedure until I had marked three other males with dots of phthalocyanine blue and quinacridone magenta and titanium white, respectively.

I was in business. Now I had an excuse, particularly on the weekends, to wander out to the eggplants at intervals throughout the day to see who was hanging out under the leaves. Over the past several weekends of *Cerceris* censusing, I have resighted marked males while adding others to the list. Of the fifteen paint-daubed wasps, eleven have returned on at least one day after receiving their dot of Liquitex. White Dot set the record, with twenty-five days between first marking and last sighting. Some males obviously have real staying power as they monitor their presumptive rendezvous site for females. I say presumptive, because I haven't seen any matings, but I am utterly confident that female *Cerceris* occasionally appear, no doubt for a brief and unspectacular copulation. To document this point, I would have to be dedicated in the extreme, willing to sit for hours in the daunting heat, ideally under the shade of a sun umbrella. I am not that dedicated. Moreover, my reputation as a neighborhood eccentric has already been firmly consolidated. No need to overdo it by entertaining passersby with a prolonged stint as a yard mannequin crouched under an umbrella, notebook in hand.

No, I am content to have learned that males of this wasp cruise among my eggplants, appearing from somewhere to go to work in the late morning and staying on through the hottest part of the day before disappearing in the late afternoon. While on duty, they are not highly aggressive about a particular perch territory. Since as many as five or six males occupy the two plants during midday, any male that tried to monopolize a large area would be engaged in a continuous battle with his rivals, not an energetically feasible or sensible investment of the male's time and energy.

Instead the males appear to be interested in waiting, and waiting some more, with regular searches through the plants to inspect whoever happens to be there, usually other patient males but occasionally a would-be mate (I guess). My eggplants probably serve only as a visually conspicuous marker, a kind of singles' bar, where females can meet males, before they go off to nest elsewhere. No females forage on the eggplants' flowers, which is, after all, an introduced species not present during the evolution of this species of *Cerceris*. Thus, it is hardly surprising that females do not use the plant's nectar. Moreover, I have not found a single nest of the wasp in or around the eggplants, suggesting that mated females do not linger near their rendezvous site.

Males of many other bees and wasps also gather at or patrol landmarks of various sorts, ranging from prominent stalks of grass to shrubs and trees growing on mountaintops. As a matter of fact, a male of a cute little gray bee, another species new to me this summer, has a patrol route that takes him to my pair of eggplants every minute or so. I catch him too, give him a white dot, and follow him on his rounds. He keeps going in circles, traveling along one row of Yellow Pear tomatoes over to the Texas sage along the split rail fence, then back along the other row of tomatoes, and on to see what's happening at the eggplants.

I'll bet that the little bee and my *Cerceris* share something in common with most other insects that have similar landmark mating systems— namely, females that scatter their burrows all over the lot rather than coming together with many others to form a little village of nesting individuals. Males of bees and wasps with group-nesting females typically patrol the nesting area, looking for the virgin females concentrated in these places. When nests are dispersed, however, finding a newly adult female emerging from a burrow is like finding the proverbial needle in a haystack. Which is where landmarks come in, as facilitators of contacts between the sexes, provided that they share similar perceptions about what makes a landmark conspicuous and attractive. I find the large, dark green leaves of my eggplants attractive and am glad that the new bee and the new wasp on the block agree.

ALTHOUGH PLEASED TO be learning new things about some newcomers to the yard and garden, I am positively thrilled to find an old friend back on my property again, the sleeping bee *Idiomelissodes duplocincta*. On the evening of May 16, weeks before I expected to see them again, males showed up on the big brittlebush in the center of the front yard, the same plant they had slept in the year before. With this discovery, I kicked my sleeping bee research program into high gear. For more than two months, I censused the bees every night except for twelve days when an East Coast trip kept me from my charges. During my absence, I authorized my wife, indeed I begged her, to fill in for me, which she did with assistance from my son Nick. As a result, throughout the summer one or another member of the family bee team headed to the front yard around the time the sun sank beneath the neighbor's roofline to the west.

Summer evenings bring neighbors out to move the sprinklers around or walk the dog. Some wondered whether I was watching the brittlebush grow or had lost my keys in the front yard. Others, having been brought up to date on sleeping bee science, smiled benignly as they passed, giving me and the bees an evening benediction.

Thanks to persistent counting, I know that the sleeping party of 46 males on May 16 grew to 318 on May 31, the high-water mark this summer. By the start of July, the group had declined to under 100, falling to 36 on the 10th of this month. Last night, July 15, a mere 3 bees slept on a dried brittlebush stem together.

During their time with us, I repeated some perch-shifting experiments. Sometimes the bees followed a favorite perch to its new location, a gratifying result, but sometimes they did not, a confusing result. I tried another experiment, placing paper bags over the previous evening's popular locations. Finding their old stems unavailable, the bees settle elsewhere in the plant. On succeeding evenings, they swarm onto their new digs, having formed an attachment to the new stems, further evidence for the role odors may play in aggregating the males.

In addition to these experiments, I spend some time simply watching the bees, especially when *Apiomerus* assassin bugs show up, which they do from time to time. As assassins go, *Apiomerus* rank as amateurs in many respects. Instead of a stealthy approach, the bugs often fly directly at a perched male, attempting to crash into the bee and catch him at the same time. The technique leaves much to be desired, since bees assaulted this way almost always fly out of harm's way or, if grasped, break free before stabbed. Fortunately for *Apiomerus*, the bees apparently have no conception of how dangerous an assassin bug is. Obtuse individuals blunder in to land on a leaf or a stem right next to the predator. It's all I can do to keep from interfering with natural selection as the bug merely leans forward and snatches the dummy from its perch.

Over the summer, I record how assassin bugs capture bees, collecting information that helps me sort out the various explanations for why *Idiomelissodes* sleep together. During attacks that result in sixteen dead bees, I never see the survivors turn en masse on the bug and attempt to injure it or drive it away. Indeed, as noted, the bees either ignore the bug or may even be attracted to it. So much for the hypothesis that sleeping bees form a group defense force against predators.

Moreover, the bees rarely reacted to the escape of an attacked companion by flight from a clear and present danger. When the bug clumsily flew at clusters of bees, they did fly up as a group, but not because one alarmed bee alerted the others—rather because they all saw or felt the impact of the bug, and took appropriate personal action. When the bug approached circumspectly, walking slowly up a stem toward a victim instead of flying right at its prey, the bees paid little attention when the bug hauled one of their number off the stem. In thirteen of sixteen lethal attacks, the captured bee's companions remained on their sleeping perch, blind to the fate of their companion.

The elimination of the group-defense and many-eyes hypotheses leaves the possibility that membership in a sleeping club dilutes the risk of death for individuals. In keeping with this hypothesis, the bee clusters did not attract large gangs of assassin bugs. Indeed, no more than two bugs visited the brittlebush bees on any given night; furthermore, the assassins usually assassinated just one prey per evening, occasionally two. Since the bees weigh about a third as much as their consumer, they constitute a substantial meal, one that takes an hour or two to polish off. Given this kind of solitary, slow-eating predator, the dilution effect works reasonably well. In a cluster of 50 bees, the chance for any individual of being grabbed and killed during the evening or night is 1 or 2 out of 50. In a cluster of 100, the probability of winding up in the cold sticky arms of an assassin bug falls by a half, no consolation for the unlucky one or two bees, but reason for males to assemble in as large groups as possible.

How nice of the bees and assassin bugs to come back to explain why the bees sleep together. They have given me a distraction throughout this hard-to-bear summer, providing a reason to await each evening with pleasant expectation. It is true that I do not embrace every summer insect with the same enthusiasm I lavish on *Idiomelissodes* and *Apiomerus*. The native fire ants have established a large colony in one corner of the garden. You walk past them in the morning at your own risk for they are quick to take offense and let you have it with their sting, not quite in the honeybee category but unpleasant enough to focus your mind on ants immediately. In addition, a horrible black aphid has taken over my asparagus bean vines, coating leaves in a thin fur of aphid bodies. Hosing them off the plant provides some satisfaction, but only temporary relief for the plants, given aphid powers of reproduction. And once again the whiteflies flutter through town by the billions, eddying out in waves from the battered zucchinis when I lift a leaf to check the last squash of summer. I am tempted to get out the vacuum cleaner, per the advice of the *Republic*'s exterminator journalist.

But these few nasties are a small price to pay for the entomology club, with its brilliantly colored assassins, sleepy bees, perky wasps, trilling cicadas (just starting their summer choruses), delicate antlions (a web-winged adult flitted from the citrus this morning), and a small army of other entertainers who make the summer move along more briskly than otherwise.

Meanwhile here in greater Phoenix, people multiply like true bugs, swarming over the desert, consuming resources at a great rate, generating spectacular quantities of ozone and carbon monoxide, and demanding new golf courses and homes with cathedral ceilings. The *Arizona Republic* carries occasional letters to the editor on urban sprawl and the effects of booming growth on the desert around Phoenix, letters filled with good advice that will be ignored, letters laden with dismay.

It's enough to send me back to the garden. I hurry there to pull up a few summer weeds foolhardy enough to invade my vegetables' sacred domain. I dream of the cool fall and cooler winter days when I will scan for cabbage

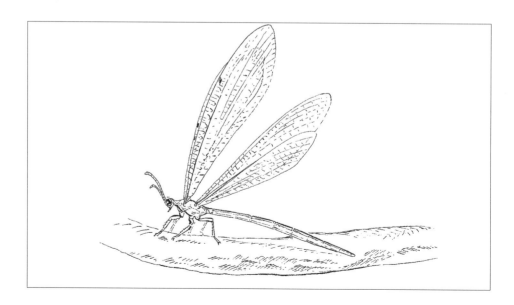

Antlion adult

loopers in a full-fledged garden. Untroubled by oppressive heat, I will check whether the second cabbage in the fifth row of my ten-row garden is ready for harvest. Happy days lie ahead when a cabbage or bowlful of spinach verges on maturity. On those happy days I will alert Sue to the impending arrival of each item. My wife has an uncanny ability to sense, unconsciously I am sure, when I am about to harvest a garden cabbage or a fresh head of lettuce, which somehow stimulates her to head to the Safeway, there to purchase cabbage or lettuce. Harsh words have been spoken when my personal produce proved redundant, a bit player on the kitchen stage, instead of honored prima donna. In the interests of marital harmony, I have learned to make sure that my cabbages do not suffer the indignity of playing second fiddle to a mass-produced, non–organically grown, store-bought cousin.

An elderly gentleman, a distant neighbor, creaks his way across the street, propped up with his cane. He wants to know which varieties of tomatoes I have planted that are now turning turtle in the sun. We discuss the relative merits of Early Girl and Celebrity, two desert-adapted types. He speaks wistfully of the days when he could bend down and work a garden. I urge him to consider a chest-high planter of some sort. We agree that watching vegetables grow is one of the most rewarding things a person can do. "It's in our genes to garden," he exclaims, before turning to walk slowly back across the street.

Cabbage suitable for harvesting

I have forgotten all about the crush of people, subsidized golf courses, and urban sprawl. What a pleasure to have a desert retreat, a place where I can find common ground with neighbors I don't really know, a place to admire the roundness of a cabbage head and its surrounding rosette of green leafy petals, or the cheerful flowers of a brittlebush, knowing that I have created a refuge not just for me but for the earwig and the sleeping bee, the desert termite and the mayfly, even the milkweed aphid and the dung caterpillar, good company one and all.

ACKNOWLEDGMENTS

In tracking down interesting papers on insect behavior, I have received help from many biologists, among them Tom Bellows, David Byrne, Thomas Eisner, Dan Gerling, Larry Hurd, Nancy Moran, and Frederick Prete. I thank Howard Evans for long ago opening my eyes to the joys of watching insects, my wife, Sue, for tolerating my entomological enthusiasms, and my parents for introducing me to gardening and encouraging my interest in compost heaps. I deeply appreciate the efforts of Sandra Dijkstra and her colleagues to improve the manuscript and find a publisher for it. My editor at W. W. Norton, Edwin Barber, gently steered me in the right direction as I revised what I had written. Turid Forsyth produced the superb line drawings for the book, for which I am grateful. Thanks to you all.

BIBLIOGRAPHY

General Works

Evans, H. E. 1984. *Life on a Little-Known Planet*. University of Chicago Press, Chicago.

Gullan, P. J., and P. S. Cranston. 1994. *The Insects: An Outline of Entomology*. Chapman & Hall, London.

Thornhill, R., and J. Alcock. 1983. *The Evolution of Insect Mating Systems*. Harvard University Press, Cambridge, MA.

2: The Gardener's Friends

Gould, S. J. 1984. "Only his wings remained." *Natural History* 93(Sept):10–18.

Graham, F., Jr. 1984. *The Dragon Hunters*. E. P. Dutton, New York.

Hurd, L. E., R. M. Eisenberg, W. F. Fagan, K. J. Tilmon, W. E. Snyder, K. S. Vandersall, S. G. Datz, and J. D. Welch. 1994. "Cannibalism reverses male-biased sex ratio in adult mantids: female strategy against food limitation?" *Oikos* 69:193–198.

Kock, A., A.-K. Jakobs, and K. Kral. 1993. "Visual prey discrimination in monocular and binocular praying mantis *Tenodera sinensis* during postembryonic development." *Journal of Insect Physiology* 39:485–491.

Kynaston, S. E., P. McErlain-Ward, and P. J. Mills. 1994. "Courtship, mating behaviour and sexual cannibalism in the praying mantis, *Sphodromantis lineola*." *Animal Behaviour* 47:739–741.

Lawrence, S. E. 1992. "Sexual cannibalism in the praying mantis, *Mantis religiosa*: a field study." *Animal Behaviour* 43:569–584.

Liske, E., and W. J. Davis. 1987. "Courtship and mating behaviour of the Chinese praying mantis, *Tenodera aridifolia sinensis*." *Animal Behaviour* 35:1524–1537.

Ordish, G. 1967. *Biological Methods in Crop Pest Control*. Constable, London.

Philbrick, H., and J. Philbrick. 1974. *The Bug Book: Harmless Insect Controls*. Garden Way Publishing, Charlotte, VT.

Prete, F. R. 1995. "Designing behavior: a case study." *Perspectives in Ethology* 11:255–277.

Roeder, K. D. 1963. *Nerve Cells and Insect Behavior*. Harvard University Press, Cambridge, MA.

3: Compost Lovers

Alcock, J. 1991. "Adaptive mate-guarding by males of *Ontholestes cingulatus* (Coleoptera: Staphylinidae)." *Journal of Insect Behavior* 4:763–771.

Eisner, T. 1960. "Defense mechanisms of arthropods. II. The chemical and mechanical weapons of an earwig." *Psyche* 60:62–70.

Lamb, R. J., and W. G. Wellington. 1975. "Life history and population characteristics of the European earwig, *Forficula auricularia* (Dermaptera: Forficulidae), at Vancouver, British Columbia." *Canadian Entomologist* 107:819–824.

Moore, A. J., and P. Wilson. 1993. "The evolution of sexually dimorphic earwig forceps: social interactions among males of the toothed earwig, *Vostox apicedentatus*." *Behavioral Ecology* 4:40–48.

Radesäter, T., and H. Halldórsdóttir. 1993. "Two male types of the common earwig: male-male competition and mating success." *Ethology* 95:89–96.

Radl, R. C., and K. E. Linsenmair. 1991. "Maternal behaviour and nest recognition in the subsocial earwig *Labidura riparia* Pallas (Dermaptera: Labiduridae)." *Ethology* 89:287–296.

Vancassel, M. 1984. "Plasticity and adaptive radiation of dermapteran parental behavior: results and perspectives." *Advances in the Study of Behavior* 14:51–79.

4: Lawn Lovers

Alcock, J., C. E. Jones, and S. L. Buchmann. 1977. "Male mating strategies in the bee *Centris pallida* Fox (Hymenoptera: Anthophoridae)." *American Naturalist* 111:145–155.

Haverty, M. I., and W. L. Nutting. 1975. "Natural wood preferences of desert termites." *American Midland Naturalist* 95:20–27.

Jones, S. C. 1990. "Colony size of the desert subterranean termite *Heterotermes aureus* (Isoptera: Rhinotermitidae)." *Southwestern Naturalist* 35:285–291.

Nutting, W. L. 1969. "Flight and colony formation." In *The Biology of Termites*, K. Krishna and F. M. Weesner, eds. Academic Press, New York.

Schaefer, D. A., and W. G. Whitford. 1981. "Nutrient cycling by the subterranean termite *Gnathamitermes tubiformans* in a Chihuahuan Desert ecosystem." *Oecologia* 48:277–283.

5: Native Stingers

Anderson, E. 1952. *Plants, Man and Life*. University of California Press, Berkeley, CA.

Evans, H. E., and M. J. West Eberhard. 1970. *The Wasps*. University of Michigan Press, Ann Arbor, MI.

Hurd, P. D., Jr., E. G. Linsley, and T. W. Whitaker. 1971. "Squash and gourd bees (*Peponapis, Xenoglossa*) and the origin of the cultivated *Cucurbita.*" *Evolution* 25:218–234.

Linsley, E. G. 1962. "Sleeping aggregations of aculeate Hymenoptera—II." *Annals of the Entomological Society of America* 55:148–164.

Nonacs, P., and H. K. Reeve. 1995. "The ecology of cooperation in wasps: causes and consequences of alternative reproductive decisions." *Ecology* 76:953–967.

Schmidt, J. O. 1990. "Hymenopteran venoms: striving toward the ultimate defense against vertebrates." In *Insect Defenses: Adaptive Mechanisms and Strategies of Prey and Predators*, D. L. Evans and J. O. Schmidt, eds. State University of New York, Albany.

Whitaker, T. W., and W. P. Bemis. 1976. "Cucurbits." In *Evolution of Crop Plants*, N. W. Simmons, ed. Longman, London.

Zavortink, T. J. 1975. "Host plants, behavior, and distribution of the eucerine bees *Idiomelissodes duplocincta* (Cockerell) and *Syntrichalonia exquisita* (Cresson)." *Pan-Pacific Entomologist* 51:236–240.

6: Camouflage Experts

Cox, G. W., and D. G. Cox. 1974. "Substrate color matching in the grasshopper *Circotettix rabula* (Orthoptera: Acrididae)." *Great Basin Naturalist* 34:60–70.

Damman, H. 1986. "The osmaterial glands of the swallowtail butterfly *Eurytides marcellus* as a defence against natural enemies." *Ecological Entomology* 11:261–265.

Eisner, T., E. van Tassell, and J. E. Carrel. 1967. "Defensive use of a 'fecal shield' by a beetle larva." *Science* 158:1471–1473.

Gillis, J. E. 1982. "Substrate colour-matching cues in the cryptic grasshopper *Circotettix rabula rabula* (Rehn & Hebard)." *Animal Behaviour* 30:113–116.

Heinrich, B. 1979. "Foraging strategies of caterpillars: leaf damage and possible predator avoidance strategies." *Oecologia* 42:325–337.

Heinrich, B., and S. L. Collins. 1983. "Caterpillar leaf damage, and the game of hide-and-seek with birds." *Ecology* 64:592–602.

Lewington, A. 1990. *Plants for People*. Oxford University Press, New York.

Nentwig, W. 1985. "A tropical caterpillar mimics faeces, leaves and a snake (Lepidoptera: Oxytenidae: *Oxytenis naemia*)." *Journal of Research on the Lepidoptera* 24:135–141.

Preston-Mafham, R., and K. Preston-Mafham. 1993. *The Encyclopedia of Land Invertebrate Behaviour*. MIT Press, Cambridge, MA.

Tinbergen, N. 1958. *Curious Naturalists*. Basic Books, New York.

7: Aliens

Blackmer, J. L., and D. N. Byrne. 1993. "Flight behavior of *Bemisia tabaci* in a vertical flight chamber: effect of time of day, sex, age, and host quality." *Physiological Entomology* 18:223–232.

Brown, J. K., D. R. Frolich, and R. C. Russell. 1995. "The sweetpotato or silverleaf whiteflies: biotypes of *Bemisia tabaci* or a species complex?" *Annual Review of Entomology* 40:511–534.

Byrne, D. N., and T. S. Bellows, Jr. 1991. "Whitefly biology." *Annual Review of Entomology* 36:431–457.

Collins, F. H., and S. M. Paskewitz. 1995. "Malaria: current and future prospects for control." *Annual Review of Entomology* 40:195–219.

Koeniger, G. 1986. "Mating sign and multiple mating in the honeybee." *Bee World* 67:141–150.

Li, T., S. B. Vinson, and D. Gerling. 1989. "Courtship and mating behavior of *Bemisia tabaci* (Homoptera: Aleyrodidae)." *Environmental Entomology* 18:800–806.

McAuliffe, J. R. 1995. "The aftermath of wildfire in the Sonoran Desert." *Sonoran Quarterly* 49(3):4–8.

Perring, T. M., A. D. Cooper, R. J. Rodriguez, C. H. Farrar, and T. S. Bellows, Jr. 1993. "Identification of a white-fly species by genomic and behavioral studies." *Science* 259:74–77.

Wenner, A. M., and W. W. Bushing. 1996. "*Varroa* mite spread in the United States." *Bee Culture* 124:341–343.

Winston, M. L. 1992. "The biology and management of Africanized honey bees." *Annual Review of Entomology* 37:173–193.

Winston, M. L. 1992. *Killer Bees: The Africanized Honey Bee in the Americas.* Harvard University Press, Cambridge, MA.

8: Transients

Edmunds, G. F., Jr., and C. H. Edmunds. 1980. "Predation, climate, and emergence and mating of mayflies." In *Advances in Ephemeroptera Biology*, J. F. Flannagan and K. E. Marshall, eds. Plenum Press, New York.

Johnson, C. G. 1969. *Migration and Dispersal of Insects in Flight.* Methuen, London.

Malcolm, S. B. 1987. "Monarch butterfly migration in North America: controversy and conservation." *Trends in Ecology and Evolution* 2:135–139.

Sauer, D., and D. Feir. 1973. "Studies on natural populations of *Oncopeltus fasciatus* (Dallas), the large milkweed bug." *American Midland Naturalist* 90:13–37.

Scudder, G. G. E., and J. Meredith. 1982. "Morphological basis of cardiac glycoside sequestration by *Oncopeltus fasciatus* (Dallas) (Hemiptera: Lygaeidae)." *Zoomorphology* 99:87–101.

Sweeney, B. W., and R. L. Vannote. 1982. "Population synchrony in mayflies: a predator satiation hypothesis." *Evolution* 36:810–821.

9: Parthenogens and Poison Eaters

Aoki, S. 1977. "*Colophina clematis* (Homoptera, Pemphigidae), an aphid species with 'soldiers.'" *Kontyu* 45:333–337.

Brower, L. P. 1984. "Chemical defense in butterflies." In *The Biology of Butterflies*, R. I. Vane-Wright and P. R. Ackery, eds. Academic Press, London.

Dussourd, D. E. 1995. "Entrapment of aphids and whiteflies in lettuce latex." *Annals of the Entomological Society of America* 88:163–172.

Dussourd, D. E., and R. F. Denno. 1991. "Deactivation of plant defense: correspondence between insect behavior and secretory canal architecture." *Ecology* 72:1383–1396.

Dussourd, D. E., and R. F. Denno. 1994. "Host range of generalist caterpillars: trenching permits feeding on plants with secretory canals." *Ecology* 75:69–78.

Dussourd, D. E., and T. Eisner. 1987. "Vein-cutting behavior: insect counterploy to the latex defense of plants." *Science* 237:898–901.

Moran, N. A. 1986. "Benefits of host plant specificity in *Uroleucon* (Homoptera: Aphididae)." *Ecology* 67:108–115.

Moran, N. A. 1992. "The evolution of aphid life cycles." *Annual Review of Entomology* 37:321–348.

Moran, N. A. 1993. "Defenders in the North American aphid *Pemphigus obesinymphae.*" *Insectes Sociaux* 40:391–402.

Moran, N. A., and P. Baumann. 1994. "Phylogenetics of cytoplasmically inherited microorganisms of arthropods." *Trends in Ecology and Evolution* 9:15–20.

Oldroyd, H. 1964. *The Natural History of Flies.* W. W. Norton, New York.

Stern, D. L., and W. A. Foster. 1996. "The evolution of soldiers in aphids." *Biological Reviews* 71:27–79.

Stern, D. L., and W. A. Foster. 1996. "The evolution of sociality in aphids: a clone's-eye view." In *Social Competition and Cooperation in Insects and Arachnids: II. Evolution of Sociality*, J. Choe and B. Crespi, eds. Cambridge University Press, Cambridge.

APPENDIX

List of Common and Scientific Names of
North American Desert Plants in Front Yard

Arizona Fairy Duster*	*Calliandra eriophylla*
Baja Fairy Duster	*Calliandra californica*
Bladderpod Mustard*	*Lesquerella gordonii*
Bluebells*	*Phacelia* spp.
Brittlebush*	*Encelia farinosa*
Candelilla	*Euphorbia antisyphilitica*
Chuparosa	*Justicia californica*
Creosote Bush	*Larrea tridentata*
Desert Marigold*	*Baileya multiradiata*
Desert Milkweed	*Asclepias subulata*
Desert Senna*	*Senna covesii*
Fiddleneck*	*Amsinckia* sp.
Firecracker Penstemon	*Penstemon eatonii*
Foothills Paloverde	*Cercidium microphyllum*
Globe Mallow*	*Sphaeralcea ambigua*
Goldeneye*	*Viguiera parishii*
Hedgehog Cactus	*Echinocereus engelmanni*
Ironwood	*Olynea tesota*
Lupine*	*Lupinus* sp.
Mexican Poppy*	*Eschscholtzia mexicana*
Owl-Clover*	*Orthocarpus purpurascens*
Parry's Penstemon*	*Penstemon parryi*
Red Hesperaloe	*Hesperaloe parviflora*
Ruellia	*Ruellia peninsularis*
Texas Ranger	*Leucophyllum frutescens*
Triangle-leaf Bursage	*Ambrosia deltoidea*
Tufted Evening Primrose	*Oenothera caespitosa*
Woolly Butterfly Bush	*Buddleia marrubiflora*

*Established from seed.

INDEX

Page numbers in *italics* refer to illustrations.